# Fundamentals of Ionizing Radiation Dosimetry

*Solutions to Exercises*

# Fundamentals of Ionizing Radiation Dosimetry

## Dosimetry

Solutions to Exercises

*Pedro Andreo, David T. Burns,*
*Alan E. Nahum, and Jan Seuntjens*

**The Authors**

*Prof. Pedro Andreo, FInstP, CPhys*
Karolinska University Hospital
171 76 Stockholm
Sweden

*Dr. David T. Burns, FInstP*
Bureau International des
Poids et Mésures
Pavillon de Breteuil
92312 Sévres Cedex
France

*Prof. Alan E. Nahum, FIPEM*
Visiting Professor
Physics Department
University of Liverpool
United Kingdom

*Prof. Jan Seuntjens, FCCPM, FAAPM, FCOMP*
McGill University
Medical Physics Unit
Cancer Research Program
Research Institute
McGill University Health Centre
1001 Décarie Blvd
Montreal QC H4A 3J1
Canada

**Cover Credits**
The cover image was kindly provided by Dr Jörg Wulff.

■ All books published by **Wiley-VCH** are carefully produced. Nevertheless, authors, editors, and publisher do not warrant the information contained in these books, including this book, to be free of errors. Readers are advised to keep in mind that statements, data, illustrations, procedural details or other items may inadvertently be inaccurate.

**Library of Congress Card No.:**
applied for

**British Library Cataloguing-in-Publication Data**
A catalogue record for this book is available from the British Library.

**Bibliographic information published by the Deutsche Nationalbibliothek**
The Deutsche Nationalbiblio-thek lists this publication in the Deutsche Nationalbibliografie; detailed bibliographic data are available on the Internet at <http://dnb.d-nb.de.>

© 2017 Wiley-VCH Verlag GmbH & Co. KGaA, Boschstr. 12, 69469 Weinheim, Germany

**Print ISBN:** 978-3-527-34352-2
**ePDF ISBN:** 978-3-527-81106-9
**ePub ISBN:** 978-3-527-81104-5
**Mobi ISBN:** 978-3-527-81105-2

**Typesetting** SPi Global, Chennai, India

# Contents

# Preface

The first edition of Frank Herbert Attix's widely used book *Introduction to Radiological Physics and Radiation Dosimetry* was published in 1986 and reprinted in 2004. The exercises and solutions at the end of each chapter were widely regarded as a useful complement to the theory described in the different chapters.

In the second edition of the book, which we abbreviate as FIORD (from *Fundamentals of IOnizing Radiation Dosimetry*), the exercises have been updated and new ones prepared for chapters and topics that were not included in the first edition. Publishing the solutions to the exercises as a separate book was considered to be a more convenient approach, and they are presented here in order to complement some of the (sometimes limited) discussions in the textbook, supported by references to its equations and figures. Hopefully, they will also be a source of inspiration for teachers to prepare new exercises.

The electronic Data Tables of the textbook, necessary for the solution of the exercises, are available from http://www.wiley-vch.de/ISBN9783527409211; physical constants and atomic data are also given in Appendix A of the textbook.

6th December 2016

*Pedro Andreo*
*David T. Burns*
*Alan E. Nahum*
*Jan Seuntjens*

# 1

# Background and Essentials

1   What is the photon energy range corresponding to the UV radiation band?
*Answer: 10 nm–400 nm corresponds to 124 eV–3.1 eV.*

**Solution:**
The quantum energy $k$ of any electromagnetic photon is given in keV by

$$k = h\nu = \frac{hc}{\lambda} = \frac{12.3982 \text{ keV Å}}{\lambda} = \frac{1.23982 \text{ keV nm}}{\lambda}$$

where $1 \text{ Å}(\text{Angstrom}) = 10^{-10}$ m, Planck's constant is $h = 6.62607 \times 10^{-34}$ J s $= 4.13561 \times 10^{-18}$ keV s (note that $1.6022 \times 10^{-16}$ J $= 1$ keV), and the velocity of light in vacuum is $c = 2.99792 \times 10^{8}$ m/s $= 2.99792 \times 10^{18}$ Å/s $= 2.99792 \times 10^{17}$ nm s$^{-1}$.

Therefore for the UV radiation, which is in the range of 10 nm–400 nm, the equation yields 124 eV–3.1 eV.

2   For a kinetic energy of 100 MeV, calculate the velocity $\beta$, for (a) electrons, (b) protons, and (c) alpha particles. The corresponding rest energies are given in the Data Tables.
*Answer: (a) 0.9999; (b) 0.4282; (c) 0.2271*

**Solution:**
We can apply either of the relations

$$\beta^2 = \frac{\tau(\tau + 2)}{(\tau + 1)^2}, \quad \text{with } \tau = E/m_0 c^2$$

or

$$\beta^2 = \frac{E(E + 2m_0 c^2)}{(E + m_0 c^2)^2}$$

From the Data Tables, the rest energies are $m_e c^2 = 0.51099$ MeV, $m_p c^2 = 938.272$ MeV, and $m_\alpha c^2 = 3727.38$ MeV. These yield
(a) Electrons: 0.9999
(b) Protons: 0.4282
(c) Alpha particles: 0.2271

*Fundamentals of Ionizing Radiation Dosimetry: Solutions to Exercises,* First Edition.
Pedro Andreo, David T. Burns, Alan E. Nahum, and Jan Seuntjens.
© 2017 Wiley-VCH Verlag GmbH & Co. KGaA. Published 2017 by Wiley-VCH Verlag GmbH & Co. KGaA.

**3** Conversely given a value of $\beta = 0.95$, calculate the corresponding kinetic energies of electrons, protons, and $\alpha$ particles.
*Answer: (a) 1.1255 MeV; (b) 2066.6 MeV; (c) 8209.86 MeV*

**Solution:**
The relation between the kinetic energy and the speed $(\beta)$ is

$$E = \frac{m_0 c^2 \beta^2}{2\sqrt{1 - \beta^2}}$$

Using the rest energies from the previous exercise, we get
(a) Electrons: 1.1255 MeV
(b) Protons: 2066.6 MeV
(c) $\alpha$-particles: 8209.86 MeV

**4** The result of a given process is derived as the product of several independent quantities, $Q = \prod q_i$. The type A and B uncertainties of each $q_i$, $(u_A, u_B)_i$, given as a relative standard uncertainty, are (0.1, 0.5), (0.01, 0.1), (0.02, 0.4), and (0.3, 0.19). Determine the combined standard uncertainty of $Q$.
*Answer: $u_c(Q) = 0.75$*

**Solution:**
Use the *law of propagation of uncertainty* twice: first for each of the respective types of uncertainty to yield the overall $u_A$ and $u_B$ types,

$$u_A = \sqrt{\sum_i u_{A_i}^2}, \quad u_B = \sqrt{\sum_i u_{B_i}^2}$$

and then for the combination of these two to yield $u_c(Q)$.
Hence

| Quantity | Rel standard uncertainty | |
|---|---|---|
| | $(u_A)_i$ | $(u_B)_i$ |
| $q_1$ | 0.10 | 0.50 |
| $q_2$ | 0.01 | 0.10 |
| $q_3$ | 0.02 | 0.40 |
| $q_4$ | 0.30 | 0.19 |
| Combined | $u_A = 0.32$ | $u_B = 0.68$ |

resulting in a combined uncertainty

$$u_c(Q) = \sqrt{u_A^2 + u_B^2} = 0.75$$

**5** Given the following set of data (75.4, 79.7, 75.0, 77.0, 78.4), with standard uncertainties (0.95, 0.5, 0.2, 1.2, 0.8), (a) determine the non-weighted and weighted means and the corresponding type A uncertainties. (b) Determine the Birge ratio for the data and comment on the uncertainty estimates of the data.
*Answer: $\bar{x} = 77.1$, $s_{\bar{x}} = 0.89$; $\bar{x}_w = 75.8$, $s_{\bar{x}_w} = 0.18$; $R_{Birge} = 2.2$*

**Solution:**

(a) Requires the straightforward application of Eqs. (1.41)–(1.46), where the different terms are

| $i$ | $x_i$ | $(x_i - \bar{x})^2$ | $s_i$ | $w_i(= 1/s_i^2)$ | $w_i\, x_i$ | $w_i(x_i - \bar{x}_w)^2$ |
|---|---|---|---|---|---|---|
| 1 | 75.4 | 2.89 | 0.95 | 1.11 | 83.55 | 0.18 |
| 2 | 79.7 | 6.76 | 0.50 | 4.00 | 318.80 | 60.79 |
| 3 | 75.0 | 4.41 | 0.20 | 25.00 | 1875.00 | 16.07 |
| 4 | 77.0 | 0.01 | 1.20 | 0.69 | 53.47 | 1.00 |
| 5 | 78.4 | 1.69 | 0.80 | 1.56 | 122.50 | 10.55 |
| $n$ | $\sum_i x_i$ | $\sum_i(x_i - \bar{x})^2$ | | $\sum_i w_i$ | $\sum_i w_i\, x_i$ | $\sum_i w_i(x_i - \bar{x}_w)^2$ |
| 5 | 385.5 | 15.8 | | 32.36 | 2453.32 | 88.58 |
| Eqs | (1.41) | (1.43) | | (1.45) | (1.44) | (1.46) num |
| | $\bar{x}$ | $s(\bar{x})$ | | $s(\bar{x}_w)_{\text{int}}$ | $\bar{x}_w$ | $s(\bar{x}_w)_{\text{ext}}$ |
| | 77.10 | 0.89 | | 0.18 | 75.80 | 0.83 |

(b) The Birge ratio is given by

$$R_{\text{Birge}} = \frac{s(\bar{x}_w)_{\text{int}}}{s(\bar{x}_w)_{\text{ext}}} = 2.2$$

$R_{\text{Birge}} = 2.2$ is a sign that some uncertainties have been under/over estimated. We typically think that we can make estimates at, say, the 20% level. A Birge significantly greater than 1.2 or 1.3 is a reasonable sign of under/overestimation. However, one proviso is the balance of uncertainties. One huge under/overestimate can make Birge large even if other uncertainties are properly estimated, especially for small data sets. This could be the case with data #3, where $s = 0.20$ might be an underestimation.

**6** Using the half-width of the set of data in the previous exercise, estimate the type B uncertainty assuming rectangular, triangular, and Gaussian (with $k = 2$) distributions. Which of the three is considered to be more conservative?
*Answer: $u_{B\,rect} = 1.36$, $u_{B\,trian} = 0.96$, $u_{B\,Gauss} = 1.18$; the 95% Gaussian is more conservative.*

**Solution:**
The half-width of the set of data, $[-L, +L]$, is determined as

$$L = \frac{\max(x_i) - \min(x_i)}{2} = \frac{79.7 - 75.0}{2} = 2.35$$

Hence

$$u_{B,\text{rect}} = \frac{L}{\sqrt{3}} = \frac{2.35}{1.73} = 1.36$$

$$u_{\text{B,trian}} = \frac{L}{\sqrt{6}} = \frac{2.35}{2.45} = 0.96$$

$$u_{\text{B,95\%}} = \frac{L}{2} = \frac{2.35}{2} = 1.18$$

The rectangular distribution is a special case, because in general for most data sets there is a higher probability that the true value lies nearer to the middle than at the extremes. This leaves the triangular and Gaussian ($k = 2 \rightarrow 95\%$) distributions being conceptually similar, with the 95% Gaussian being more conservative.

# 2

# Charged Particle Interactions

1  For the Rutherford (Geiger and Marsden) experiment with 5.5 MeV
α particles on a 1 μm gold foil and for the six angles (decades) between
$10^{-5}$ and $10^0$ rad, calculate the Rutherford differential cross section (DCS),
$d\sigma_R/d\Omega$, (a) without and (b) with screening. Represent both results
graphically and draw conclusions.
*Answer: The non-screened DCS values vary between $1.7137 \times 10^{-3}$ and
$2.0273 \times 10^{-23}$ cm$^2$ rad$^{-1}$ in the interval $[10^{-5}-1$ rad]. The screening angle
is $3.7 \times 10^{-3}$ rad, and the corresponding screened DCS values vary between
$9.3620 \times 10^{-14}$ and $2.0273 \times 10^{-23}$ cm$^2$ rad$^{-1}$. The screening $\chi_a$ cuts off the
otherwise increasing DCS with decreasing angle, which remains practically
constant below $\chi_a$.*

**Solution:**
Note that the foil thickness is irrelevant, as the interaction is assumed to be
with one atom of gold, $Z = 79$. The α-particle charge is $z = 2$.
(a)  The non-screened Rutherford DCS is given by

$$\frac{d\sigma_R}{d\Omega} = r_e^2 (z\,Z)^2 \left(\frac{m_e c^2}{m_0 c^2}\right)^2 \frac{1-\beta^2}{\beta^4} \frac{1}{(1-\cos\theta)^2}$$

where the relevant constants given in the Data Tables are $m_e c^2 = 0.51099$ MeV, $m_0 c^2 = m_\alpha c^2 = 3727.38$ MeV, $r_e = 2.81794 \times 10^{-13}$ cm,
and $\beta^2 = \tau(\tau+2)/(\tau+1)^2$ with $\tau = E/m_\alpha c^2$. This yields $\beta = 0.05426$.
The DCS can then be calculated; the results are given in the table below.
(b)  The screened Rutherford DCS is given by

$$\frac{d\sigma_R}{d\Omega} = (z\,Z\,r_e)^2 \left(\frac{m_e c^2}{m_0 c^2}\right)^2 \frac{1-\beta^2}{\beta^4} \frac{1}{(1-\cos\theta + 0.5\chi_a^2)^2},$$

which differs from the non-screened DCS in the screening angle $\chi_a$ in
the denominator. This is taken to be that of Molière, which is given by

$$\chi_a^2 = \chi_0^2 \left[1.13 + 3.76\left(\frac{z\,Z}{137\beta}\right)^2\right]$$

*Fundamentals of Ionizing Radiation Dosimetry: Solutions to Exercises*, First Edition.
Pedro Andreo, David T. Burns, Alan E. Nahum, and Jan Seuntjens.

where

$$\chi_0 = \frac{4.2121 \times 10^{-3} \sqrt{1 - \beta^2}}{m_0 c^2 \beta} Z^{1/3}$$

Thus the screening values obtained are

$$\chi_0 = 8.922 \times 10^{-5} \text{ rad} \quad (0.005°)$$
$$\chi_a = 3.678 \times 10^{-3} \text{ rad} \quad (0.211°),$$

which can be inserted in the screened DCS and its values obtained. Putting together the results for the two DCS, we get the following table:

| $\theta$(rad) | Rutherford DCS (cm² rad⁻¹) | |
| --- | --- | --- |
| | Non-screened | Screened |
| $10^{-5}$ | $1.7137 \times 10^{-3}$ | $9.3620 \times 10^{-14}$ |
| $10^{-4}$ | $1.7137 \times 10^{-7}$ | $9.3483 \times 10^{-14}$ |
| $10^{-3}$ | $1.7137 \times 10^{-11}$ | $8.1178 \times 10^{-14}$ |
| $10^{-2}$ | $1.7137 \times 10^{-15}$ | $1.3296 \times 10^{-15}$ |
| $10^{-1}$ | $1.7165 \times 10^{-19}$ | $1.7119 \times 10^{-19}$ |
| $10^{0}$ | $2.0273 \times 10^{-23}$ | $2.0273 \times 10^{-23}$ |

which is shown in Figure 2.1 and shows the influence of screening, which at about $4 \times 10^{-3}$ rad cuts the otherwise increasing DCS with angle, remaining practically constant below $\chi_a$.

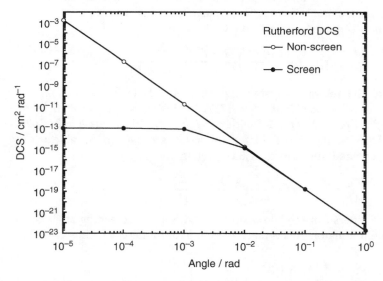

**Figure 2.1** Rutherford differential cross section, with and without screening, for 5.5 MeV α-particles on a 1 μm gold foil.

**2** Assume the Rutherford experiment to have been made with protons having the same speed ($\beta$) than the original 5.5 MeV $\alpha$ particles. Calculate the screened Rutherford DCS for the same angles as above. Compare graphically with the case of $\alpha$ particles from the previous exercise and draw conclusions.

*Answer: In this case the screening angle is $\sim 7 \times 10^{-3}$ rad, almost double than that for $\alpha$ particles with the same $\beta$, and the DCS varies between $2.3635 \times 10^{-14}$ and $7.9977 \times 10^{-23}$ in the interval $[10^{-5}-1$ rad$]$. The DCS of protons is, above the screening angle, larger than that of the $\alpha$ particles, as expected from the lower proton rest energy. The larger screening angle for protons, due also to the $1/m_0c^2$ dependence of $\chi_0$, cuts off the DCS earlier (when decreasing angle) than for $\alpha$ particles, so that for $\theta < \chi_a$, $DCS_{proton} < DCS_{alpha}$.*

**Solution:**
The constants and expressions to be used are the same as in the previous exercise, with the exception of $m_pc^2 = 938.272$ MeV and $z = 1$. As above, $\beta = 0.05426$, which corresponds to a proton energy of 1.384 MeV.
In this case the screening values are

$$\chi_0 = 3.544 \times 10^{-4} \text{ rad} \quad (0.020°)$$
$$\chi_a = 7.313 \times 10^{-3} \text{ rad} \quad (0.419°)$$

which when inserted in the expression for the screened DCS give

| | Rutherford DCS (cm² rad⁻¹) |
|---|---|
| $\theta$(rad) | Screened |
| $10^{-5}$ | $2.3635 \times 10^{-14}$ |
| $10^{-4}$ | $2.3626 \times 10^{-14}$ |
| $10^{-3}$ | $2.2775 \times 10^{-14}$ |
| $10^{-2}$ | $2.8701 \times 10^{-15}$ |
| $10^{-1}$ | $6.7005 \times 10^{-19}$ |
| $10^{0}$ | $7.9977 \times 10^{-23}$ |

Figure 2.2 compares the DCS for protons and for the $\alpha$ particles from the previous exercise. The figure shows that, above the screening angle, the DCS is larger for protons than for $\alpha$ particles, as expected from the lower proton rest energy. The larger screening angle for protons ($\sim 7.3 \times 10^{-3}$ rad) than for $\alpha$ particles ($\sim 4 \times 10^{-3}$ rad), due also to the $1/m_0c^2$ dependence of $\chi_0$, cuts off the DCS earlier (when decreasing angle) than in the alpha case, and from that point downward, the DCS is smaller for protons than for the $\alpha$ particles whenever $\theta < \chi_a$.

**3** (a) Derive an expression for the relativistic screened Rutherford total cross section (TCS) restricted to angles smaller than the atomic screening angle $\chi_a$. (b) Compare with the unrestricted expression (Eq. (2.15)) for 5 MeV

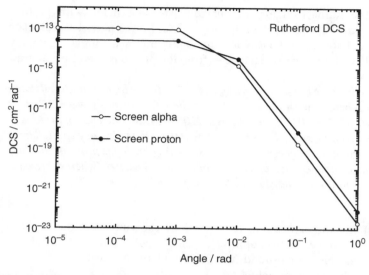

**Figure 2.2** Screened Rutherford differential cross section for protons and α particles having the same speed $\beta$, incident on a 1 μm gold foil.

electrons incident on a mercury atom. (c) Calculate the mean free path of these electrons due to elastic collisions and the subsequent number of elastic collisions per path length.

*Answer: (a)* $\sigma_R[0, \chi_a] = (z\,Z\,r_e)^2 \left(\dfrac{m_e c^2}{m_0 c^2}\right)^2 \dfrac{1-\beta^2}{\beta^4} \dfrac{8\pi(1-\cos\chi_a)}{\chi_a^2(2+\chi_a^2-2\cos\chi_a)}$.

*(b)* $\sigma_R[0, \pi] = 2.10731 \times 10^{-18} cm^2$ *and* $\sigma_R[0, \chi_a] = 1.05366 \times 10^{-18} cm^2$, *that is, elastic collisions with* $\theta \leq \chi_a$ *account for practically 50% of the TCS.*
*(c) MFP = 0.12 μm, that is, almost 10 elastic collisions occur per micron.*

**Solution:**
(a) The screened Rutherford DCS is given by

$$\frac{d\sigma_R}{d\Omega} = (z\,Z\,r_e)^2 \left(\frac{m_e c^2}{m_0 c^2}\right)^2 \frac{1-\beta^2}{\beta^4} \frac{1}{(1-\cos\theta+0.5\chi_a^2)^2}$$

where $\chi_a$ is the screening angle, which in Molière's theory is given by

$$\chi_a^2 = \frac{4.2121 \times 10^{-3}\sqrt{1-\beta^2}}{m_0 c^2 \beta} Z^{1/3} \left[1.13 + 3.76\left(\frac{z\,Z}{137\beta}\right)^2\right]$$

The screened Rutherford TCS (Eq. (2.15)) was obtained from

$$\sigma_R[0, \pi] = \int_0^\pi \frac{d\sigma_R}{d\Omega}\,d\Omega = \int_0^\pi \frac{d\sigma_R}{d\Omega} 2\pi\sin\theta\,d\theta$$

$$= (z\,Z\,r_e)^2 \left(\frac{m_e c^2}{m_0 c^2}\right)^2 \frac{1-\beta^2}{\beta^4} \frac{16\,\pi}{\chi_a^2(4+\chi_a^2)}$$

Instead of integrating in the interval $[0, \pi]$, it is convenient to do it in the interval $[\theta_{min}, \theta_{max}]$, that is,

$$\sigma_R[\theta_{min}, \theta_{max}] = \int_{\theta_{min}}^{\theta_{max}} \frac{d\sigma_R}{d\Omega} 2\pi \sin\theta \, d\theta$$

$$= (z \, Z \, r_e)^2 \left(\frac{m_e c^2}{m_0 c^2}\right)^2 \frac{1 - \beta^2}{\beta^4} 2\pi$$

$$\left[\frac{1}{1 - \cos\theta_{min} + 0.5\chi_a^2} - \frac{1}{1 - \cos\theta_{max} + 0.5\chi_a^2}\right]$$

which for $\theta_{min} = 0$ and $\theta_{max} = \pi$ coincides with the expression above. For the present exercise, we then take $\theta_{min} = 0$ and $\theta_{max} = \chi_a$ and get

$$\sigma_R[0, \chi_a] = \int_0^{\chi_a} \frac{d\sigma_R}{d\Omega} 2\pi \sin\theta \, d\theta$$

$$= (z \, Z \, r_e)^2 \left(\frac{m_e c^2}{m_0 c^2}\right)^2 \frac{1 - \beta^2}{\beta^4} \frac{8\pi(1 - \cos\chi_a)}{\chi_a^2(2 + \chi_a^2 - 2\cos\chi_a)}$$

(b) For 5 MeV electrons on mercury ($Z = 80$, $A = 200.59$, $\rho = 13.546$ g cm$^{-3}$), $\beta = 0.995692$, $m_0 c^2 = m_e c^2$. Thus $\chi_a = 5.14873 \times 10^{-3}$ rad (0.3°). Note that we could have replaced $Z^2$ by $Z(Z + 1)$. Therefore

$$\sigma_R[0, \pi] = 2.10731 \times 10^{-18} \text{ cm}^2$$
$$\sigma_R[0, \chi_a] = 1.05366 \times 10^{-18} \text{ cm}^2$$

that is, elastic collisions with $\theta \leq \chi_a$ account for practically 50% of the TCS.

The result above explains why even if we would have used the small-angle approximation,

$$\frac{d\sigma_{R,small}}{d\Omega} = (z \, Z \, r_e)^2 \frac{1 - \beta^2}{\beta^4} \frac{4}{(\theta^2 + \chi_a^2)^2}$$

with corresponding TCS

$$\sigma_{R,small}[\theta_{min}, \theta_{max}] = \int_{\theta_{min}}^{\theta_{max}} \frac{d\sigma_R}{d\Omega} 2\pi\theta \, d\theta$$

$$= (z \, Z \, r_e)^2 \frac{1 - \beta^2}{\beta^4} 4\pi \left[\frac{1}{\theta_{min}^2 + \chi_a^2} - \frac{1}{\theta_{max}^2 + \chi_a^2}\right]$$

that is, for $\theta_{max} = 1$ rad and $\theta_{max} = \chi_a$, we get, respectively,

$$\sigma_{R,small}[0, 1] = (z \, Z \, r_e)^2 \frac{1 - \beta^2}{\beta^4} \frac{4\pi}{\chi_a^2(1 + \chi_a^2)},$$

$$\sigma_{R,small}[0, \chi_a] = (z \, Z \, r_e)^2 \frac{1 - \beta^2}{\beta^4} \frac{2\pi}{\chi_a^2}$$

the results would have been practically identical, as $\chi_a$ is very small and most collisions occur within $\theta \leq \chi_a$.

(c) The mean free path is given by MFP $= A/\rho\, N_A \sigma_R[0,\pi]$, and the atom density is $n_a = \rho N_A/A = 4.0668 \times 10^{22}$ cm$^{-3}$. Together with $\sigma_R[0,\pi] = 2.10731 \times 10^{-18}$ cm$^2$, they yield MFP $= 0.11669$ μm. This means that a 5 MeV electron in mercury suffers about 10 elastic collisions per micron (!).

**4** (a) Calculate the Gaussian multiple scattering distribution (including screening) of electrons with kinetic energy $2m_e c^2$ impinging on a lead foil $2.5 \times 10^{-2}$ mm thick. (b) Determine the corresponding mass scattering power.

*Answer: (a) The distribution has the form $\exp(-\theta^2/\overline{\theta^2})/\pi\overline{\theta^2}$ and has a mean square scattering angle $\overline{\theta^2} = 0.609$ rad$^2$. (b) The mass scattering power is $T/\rho = 21.48$ rad$^2$ cm$^2$ g$^{-1}$.*

**Solution:**

(a) The Gaussian distribution for the multiple small-angle scattering theory is given by

$$F_G(\theta, t) = \frac{1}{\pi\, \overline{\theta^2}} \exp\left(-\frac{\theta^2}{\overline{\theta^2}}\right)$$

where the mean square scattering angle is given by

$$\overline{\theta^2} = 4\,\pi\,\rho\,t\frac{N_A}{A}(z\,Z\,r_e)^2\left(\frac{m_e c^2}{m_0 c^2}\right)^2 \frac{1-\beta^2}{\beta^4}\left[\ln\left(1 + \frac{\theta_{max}^2}{\theta_{min}^2}\right) - \frac{1}{1 + \frac{\theta_{min}^2}{\theta_{max}^2}}\right]$$

with

$$\theta_{min} = \chi_0 = \frac{4.2121 \times 10^{-3}\sqrt{1-\beta^2}}{m_0 c^2 \beta} Z^{1/3}$$

$$\theta_{max} = \frac{184.4\sqrt{1-\beta^2}}{m_0 c^2 \beta}\frac{1}{A^{1/3}}$$

and the constrain that $\theta_{max} \ll 1$ rad. The distribution is normalized as $\int_0^\pi 2\pi\theta F_G(\theta, t) = 1$.

For electrons of 1.022 MeV onto lead ($Z = 82, A = 207.2, \rho = 11.35$ g cm$^{-3}$), $\beta = 0.942809$, we have

$$\theta_{min} = 0.0127 \text{ rad } (0.725°)$$
$$\theta_{max} = 1 \text{ rad } (57.3°)$$

and for $t = 0.0025$ cm, the mean square scattering angle becomes $\overline{\theta^2} = 0.609$ rad$^2$ (i.e., the mean scattering angle is $\bar{\theta} = 0.78$ rad (44.7°)). We can then build the Gaussian distribution table as follows:

| $\theta$ (rad) | $\theta$ (°) | $F_G(\theta, t)$ |
|---|---|---|
| 0 | 0 | 0.522681 |
| 0.1 | 5.72958 | 0.514168 |

| $\theta$ (rad) | $\theta$ (°) | $F_G(\theta, t)$ |
|---|---|---|
| 0.2 | 11.4592 | 0.489453 |
| 0.3 | 17.1887 | 0.450873 |
| 0.4 | 22.9183 | 0.401916 |
| 0.5 | 28.6479 | 0.346699 |
| 0.6 | 34.3775 | 0.289407 |
| 0.7 | 40.107 | 0.233777 |
| 0.8 | 45.8366 | 0.182739 |
| 0.9 | 51.5662 | 0.138229 |
| 1 | 57.2958 | 0.101182 |
| 1.1 | 63.0254 | 0.0716713 |
| 1.2 | 68.7549 | 0.0491275 |
| 1.3 | 74.4845 | 0.0325868 |
| 1.4 | 80.2141 | 0.0209168 |
| 1.5 | 85.9437 | 0.0129923 |
| 1.6 | 91.6732 | 0.00780936 |
| 1.7 | 97.4028 | 0.00454236 |
| 1.8 | 103.132 | 0.00255673 |
| 1.9 | 108.862 | 0.0013926 |
| 2 | 114.592 | 0.000734013 |

whose graphical representation is shown in Figure 2.3.

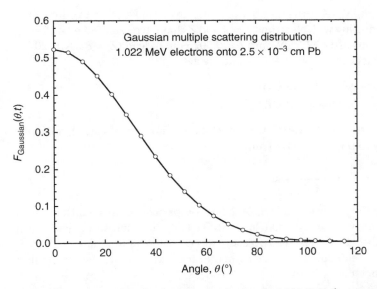

**Figure 2.3** Gaussian multiple scattering distribution for 1.022 MeV electrons onto 0.0025 cm of lead.

(b) The mass scattering power is simply given by

$$\frac{T}{\rho} = \frac{\overline{\theta^2}}{\rho t} = \frac{0.609 \text{ rad}^2}{0.02835 \text{ g cm}^{-2}} = 21.48 \text{ rad}^2 \text{ cm}^2 \text{ g}^{-1}.$$

**5** Repeat the previous exercise for the Molière multiple scattering distribution, estimating the number of collisions. Compare the full Molière distribution with that using only the first term of the series (Gaussian), and include in the comparison the Gaussian distribution from the previous exercise.

*Answer: The basic parameters are* $\chi_a = 0.020589$ *rad and* $\chi_c = 0.280648$ *rad. They yield* $\Omega_0 \approx 159$ *collisions. Molière's parameter* $B_{Mol} = 6.9707$ *yields* $\chi_c^2 B_{Mol} = 0.549$ *rad$^2$, from where the distribution* $\frac{1}{2\pi \, \chi_c^2 B_{Mol}}[f^{(0)}(\vartheta) + B_{Mol}^{-1} f^{(1)}(\vartheta) + B_{Mol}^{-2} f^{(2)}(\vartheta)]$ *can be determined using the values of the* $f^{(i)}(\vartheta)$ *functions given in Table 2.1 of the textbook.*

**Solution:**
The basic parameters entering into Molière's theory are the screening angle, $\chi_a$, and the "unit probability angle," $\chi_c$. They are given by the expressions

$$\chi_a^2 = \left(\frac{4.2121 \times 10^{-3}\sqrt{1-\beta^2}}{m_e c^2 \beta} Z^{1/3}\right)^2 \left[1.13 + 3.76 \frac{Z(Z+1)}{(137\beta)^2}\right]$$

and

$$\chi_c^2 = 4\pi r_e^2 Z(Z+1) \, n_a \frac{1-\beta^2}{\beta^4} t$$

where $n_a = \rho N_A/A$ is the number of scattering atoms per unit volume and, as we are dealing with electrons, $z = 1$ and $Z^2$ has been replaced by $Z(Z+1)$. For electrons of 1.022 MeV onto $t = 0.0025$ cm of lead ($Z = 82, A = 207.2$, $\rho = 11.35$ g cm$^{-3}$), $\beta = 0.942809$, and we get

$$\chi_a = 0.020589 \text{ rad } (1.18°)$$
$$\chi_c = 0.280648 \text{ rad } (16.08°)$$

These two enable the estimation of Molière's number of elastic collisions occurring in the thickness $t$ from

$$\Omega_0 = e^b = \frac{\chi_c^2}{1.167\chi_a^2} \approx 159$$

and yields the parameter $b \approx 5.0702$. For comparison, if we would have used the procedure of exercise #3 (Rutherford TCS), the result would have been $\sim 184$ collisions, that is, the two estimates coincide quite well.

Following the detailed steps described in the textbook, we determine first the parameter $B_{Mol}$ from

$$B_{Mol} = 1.083 + 2.719\log_{10}\Omega_0 - 0.0205(\log_{10}\Omega_0)^2 = 6.97071$$

Note that using a numerical iteration procedure we get 7.019, that is, a difference of ~0.7%, which can be considered negligible in the overall process. From this we obtain

$$\chi_c^2 B_{\mathrm{Mol}} = 0.54904 \text{ rad}^2,$$

which would correspond to the mean square scattering angle of the Gaussian term. Note that one of the conditions set by Bethe was $\chi_c^2 B_{\mathrm{Mol}} \leq 1$, which in this case is satisfied. Note also that in the previous exercise we had obtained $\overline{\theta^2} = 0.609 \text{ rad}^2$, that is, not too different. Considering now that Molière's reduced scattering angle is defined by

$$\vartheta = \frac{\theta}{\chi_c \sqrt{B_{\mathrm{Mol}}}},$$

we can calculate the angular distribution

$$F_{\mathrm{Mol}}(\theta, s) = \frac{1}{2\pi \, \chi_c^2 B_{\mathrm{Mol}}} [f^{(0)}(\vartheta) + B_{\mathrm{Mol}}^{-1} f^{(1)}(\vartheta) + B_{\mathrm{Mol}}^{-2} f^{(2)}(\vartheta)]$$

using the values of the functions $f^{(0)}(\vartheta)$(Gaussian), $f^{(1)}(\vartheta)$, and $f^{(2)}(\vartheta)$ given in Table 2.1 of the textbook. The following table is obtained:

| $\theta$ (rad) | $\theta$ (°) | $F_{\mathrm{Mol}}$ | $F_{\mathrm{Mol\text{-}G}}$ | $\vartheta$ | $f^{(0)}(\vartheta)$ | $f^{(1)}(\vartheta)$ | $f^{(2)}(\vartheta)$ |
|---|---|---|---|---|---|---|---|
| 0 | 0.00 | 6.389E−01 | 5.880E−01 | 0.000 | 2.000E+00 | 8.456E−01 | 2.493E+00 |
| 0.1 | 5.73 | 6.240E−01 | 5.772E−01 | 0.136 | 1.963E+00 | 7.786E−01 | 2.291E+00 |
| 0.2 | 11.46 | 5.818E−01 | 5.461E−01 | 0.272 | 1.858E+00 | 5.926E−01 | 1.745E+00 |
| 0.3 | 17.19 | 5.179E−01 | 4.979E−01 | 0.408 | 1.694E+00 | 3.274E−01 | 1.005E+00 |
| 0.4 | 22.92 | 4.407E−01 | 4.375E−01 | 0.544 | 1.488E+00 | 3.700E−02 | 2.623E−01 |
| 0.5 | 28.65 | 3.591E−01 | 3.705E−01 | 0.680 | 1.260E+00 | −2.244E−01 | −3.142E−01 |
| 0.6 | 34.38 | 2.810E−01 | 3.024E−01 | 0.815 | 1.029E+00 | −4.154E−01 | −6.284E−01 |
| 0.7 | 40.11 | 2.120E−01 | 2.378E−01 | 0.951 | 8.090E−01 | −5.153E−01 | −6.737E−01 |
| 0.8 | 45.84 | 1.550E−01 | 1.803E−01 | 1.087 | 6.132E−01 | −5.249E−01 | −5.172E−01 |
| 0.9 | 51.57 | 1.106E−01 | 1.317E−01 | 1.223 | 4.479E−01 | −4.623E−01 | −2.629E−01 |
| 1 | 57.30 | 7.764E−02 | 9.271E−02 | 1.359 | 3.153E−01 | −3.550E−01 | −1.134E−02 |
| 1.1 | 63.03 | 5.416E−02 | 6.290E−02 | 1.495 | 2.139E−01 | −2.312E−01 | 1.692E−01 |
| 1.2 | 68.75 | 3.782E−02 | 4.113E−02 | 1.631 | 1.399E−01 | −1.145E−01 | 2.528E−01 |
| 1.3 | 74.48 | 2.660E−02 | 2.592E−02 | 1.767 | 8.815E−02 | −1.958E−02 | 2.494E−01 |
| 1.4 | 80.21 | 1.889E−02 | 1.574E−02 | 1.903 | 5.353E−02 | 4.735E−02 | 1.896E−01 |
| 1.5 | 85.94 | 1.354E−02 | 9.211E−03 | 2.039 | 3.133E−02 | 8.697E−02 | 1.080E−01 |
| 1.6 | 91.67 | 9.790E−03 | 5.195E−03 | 2.175 | 1.767E−02 | 1.042E−01 | 3.184E−02 |
| 1.7 | 97.40 | 7.137E−03 | 2.821E−03 | 2.311 | 9.595E−03 | 1.055E−01 | −2.355E−02 |
| 1.8 | 103.13 | 5.256E−03 | 1.479E−03 | 2.446 | 5.031E−03 | 9.724E−02 | −5.469E−02 |
| 1.9 | 108.86 | 3.921E−03 | 7.467E−04 | 2.582 | 2.540E−03 | 8.442E−02 | −6.476E−02 |
| 2 | 114.59 | 2.973E−03 | 3.626E−04 | 2.718 | 1.233E−03 | 7.051E−02 | −6.084E−02 |
| 2.1 | 120.32 | 2.297E−03 | 1.703E−04 | 2.854 | 5.791E−04 | 5.747E−02 | −4.971E−02 |

| $\theta$ (rad) | $\theta$ (°) | $F_{Mol}$ | $F_{Mol\text{-}G}$ | $\vartheta$ | $f^{(0)}(\vartheta)$ | $f^{(1)}(\vartheta)$ | $f^{(2)}(\vartheta)$ |
|---|---|---|---|---|---|---|---|
| 2.2 | 126.05 | 1.809E−03 | 7.697E−05 | 2.990 | 2.618E−04 | 4.624E−02 | −3.662E−02 |
| 2.3 | 131.78 | 1.450E−03 | 3.340E−05 | 3.126 | 1.136E−04 | 3.707E−02 | −2.471E−02 |
| 2.4 | 137.51 | 1.180E−03 | 1.405E−05 | 3.262 | 4.779E−05 | 2.978E−02 | −1.528E−02 |
| 2.5 | 143.24 | 9.722E−04 | 5.692E−06 | 3.398 | 1.936E−05 | 2.410E−02 | −8.557E−03 |
| 2.6 | 148.97 | 8.090E−04 | 2.205E−06 | 3.534 | 7.499E−06 | 1.969E−02 | −4.137E−03 |
| 2.7 | 154.70 | 6.789E−04 | 8.319E−07 | 3.670 | 2.829E−06 | 1.625E−02 | −1.436E−03 |
| 2.8 | 160.43 | 5.738E−04 | 3.026E−07 | 3.806 | 1.029E−06 | 1.356E−02 | 9.142E−05 |
| 2.9 | 166.16 | 4.881E−04 | 1.048E−07 | 3.941 | 3.566E−07 | 1.142E−02 | 8.831E−04 |
| 3 | 171.89 | 4.178E−04 | 3.584E−08 | 4.077 | 1.219E−07 | 9.707E−03 | 1.235E−03 |
| 3.1 | 177.62 | 3.597E−04 | 1.209E−08 | 4.213 | 4.111E−08 | 8.317E−03 | 1.342E−03 |

where the fourth column corresponds to the Gaussian term of the series. A plot of the two distributions (solid and dashed lines), including also the Gaussian distribution from the previous exercise (dotted) is shown in Figure 2.4, which shows a substantial difference between the two Gaussians at small angles, $F_{Mol\text{-}G}$ being more peaked in the forward direction than the conventional $F_{Gaussian}$.

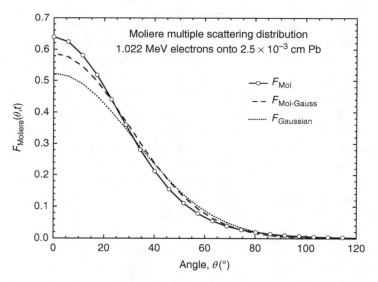

**Figure 2.4** Molière multiple scattering distribution for 1.022 MeV electrons onto 0.0025 cm of lead. The full distribution corresponds to the solid line with circles, Molière's Gaussian term to the dashed line, and the conventional Gaussian distribution to the dotted line.

**6** What is the approximate probability of a single charged particle achieving a path length equal to twice its CSDA range?
*Answer: Zero*

**Solution:**
By definition, $R_{CSDA}$ is the average path length that a charged particle travels until it has lost its energy. Therefore there is no chance that it will travel further than this distance (except for minor range straggling) and the probability is zero.

7 What is the maximum energy that can be transferred to an electron in a hard collision by a 25 MeV (a) electron, (b) positron, (c) proton, and (d) α particle?
*Answer: (a) Electron: 12.5 MeV; (b) Positron: 25 MeV; (c) Proton: 0.0552 MeV; (d) α particle: 0.0138 MeV*

**Solution:**
(a) In the case of electrons, the primary and the electron emerging after the collision are indistinguishable according to Dirac theory (exchange effect). Thus, by convention, an electron cannot lose more than half its energy in a collision with another electron, and therefore $W_{max} = 25/2 = 12.5$ MeV.
(b) A positron can lose all its energy in a collision; therefore $W_{max} = 25$ MeV.
(c) For a proton

$$W_{max}(\text{proton}) = \frac{2m_e c^2 \beta^2}{1 - \beta^2} = 1.022 \frac{\beta^2}{1 - \beta^2}$$

$$= 1.022 \frac{0.051232}{1 - 0.051232} = 0.0552 \text{ MeV}$$

(d) For an α particle

$$W_{max}(\text{alpha}) = \frac{2m_e c^2 \beta^2}{1 - \beta^2} = 1.022 \frac{\beta^2}{1 - \beta^2}$$

$$= 1.022 \frac{0.01328}{1 - 0.01328} = 0.0138 \text{ MeV}$$

8 Redo the previous exercise for the case where each of the particles has the same velocity as a 25 MeV proton.
*Answer: (a) Electron: $6.81 \times 10^{-3}$ MeV; (b) Positron: 0.01362 MeV; (c) and (d) Proton and α particle: 0.05519 MeV*

**Solution:**
Same velocity means same $\beta$. For a 25 MeV proton, from the previous exercise, $\beta^2 = 0.051232$. For electrons or positrons, the kinetic energy is

$$E = m_0 c^2 \left( \frac{1}{\sqrt{1 - \beta^2}} - 1 \right) = 0.511 \left( \frac{1}{\sqrt{1 - 0.051232}} - 1 \right) = 0.01362 \text{ MeV}$$

and therefore
(a) $W_{max} = E/2 = 6.81 \times 10^{-3}$ MeV $= 6.8$ keV

(b) $W_{max} = E = 0.01362$ MeV $= 13.6$ keV
(c) For protons

$$W_{max} = \frac{2m_e c^2 \beta^2}{1 - \beta^2} = 1.022 \frac{\beta^2}{1 - \beta^2}$$

$$= 1.022 \frac{0.051232}{1 - 0.051232} = 0.0552 \text{ MeV}$$

(d) For $\alpha$ particles, the maximum energy transfer is the same as for protons.

**9** Calculate the mass electronic stopping power for a 800 MeV triton in copper. Assume the shell, Bloch, and Barkas corrections to be negligible. Approximate the density effect using Sternheimer model with the parameters $C = -4.4190, X_0 = -0.0254, X_1 = 3.2792, a = 1.4339,$ $m = 2.9044,$ and $\delta(X_0) = 0.08$.
Answer: $S_{el}/\rho = 2.56295$ MeV cm$^2$ g$^{-1}$

**Solution:**
The expression for the mass electronic stopping power is

$$S_{el}/\rho = 4\pi r_e^2 m_e c^2 N_A \times \frac{Z}{A} \frac{z^2}{\beta^2} \times L(\beta)$$

where the constant $4\pi r_e^2 m_e c^2 N_A = 0.307074$ MeV cm$^2$ g$^{-1}$.
$L(\beta)$ is the stopping number, $L(\beta) = L_0(\beta) + z L_1(\beta) + z^2 L_2(\beta)$, and we assume $L_1(\beta)$ and $L_2(\beta)$ to be negligible. Thus the only component of importance is $L_0(\beta)$

$$L_0(\beta) = \ln \left[ \frac{2m_e c^2 \beta^2}{(1 - \beta^2)I} \right] - \beta^2 - \frac{C(\beta)}{Z} - \frac{\delta(\beta)}{2}$$

where we also assume $C(\beta)$ to be negligible (note that all the corrections mentioned are important at low energies), but the density effect needs to be calculated.
Triton has a rest energy $m_0 c^2 = 2808.921$ MeV and charge $z = 1$. For a kinetic energy of $E = 800$ MeV, $\tau = 0.2848$, and $\beta = 0.62786$.
For the copper target $Z = 29, A = 63.546, \rho = 8.96$ g cm$^{-3}$, and its $I$-value is 322 eV. The Sternheimer parameter $X$ for the density effect correction is defined as

$$X = 0.5 \log_{10}[\tau(\tau + 2)] = -0.0933$$

which is smaller than $X_0$ $(-0.0254)$, and therefore the density effect is given by

$$\delta(X) = 10^{2(X - X_0)} \delta(X_0) = 0.05852$$

Note that this is the condition for 'low energies', as the triton kinetic energy is much smaller than its rest energy.
With this density effect correction and the values of $\beta$ and $I$, $L_0(\beta) = 7.2096$ and $S_{el}/\rho = 2.56295$ MeV cm$^2$ g$^{-1}$.

**10** (a) At what kinetic energy would an $\alpha$ particle have the same velocity as the triton particle in the previous exercise? (b) What is the mass electronic stopping power of such an $\alpha$ particle also in copper?
*Answer: (a) E = 1061.58 MeV; (b) $S_{el}/\rho$ = 10.2929 MeV cm$^2$ g$^{-1}$*

**Solution:**

(a) As the same value of $\tau = E/m_0c^2$ gives the same $\beta$

$$\frac{E(\alpha)}{m_0c^2(\alpha)} = \frac{E(\text{triton})}{m_0c^2(\text{triton})}$$

$$\frac{E(\alpha)}{3727.379} = \frac{800}{2808.921} \Rightarrow E(\alpha) = 1061.58 \text{ MeV}$$

(b) As $\beta$ is the same, and the material is the same, $L_0(\beta)$ is also the same. For the $\alpha$ particle, $z^2 = 4$; thus $S_{el}/\rho = 2.56295 \times 4 = 10.2518$ MeV cm$^2$ g$^{-1}$.

**11** In a cyclotron the maximum energy is proportional to $z^2/m_0c^2$. Assume that one such machine is capable of accelerating protons to a maximum energy of 1000 MeV. (a) What are the approximate maximum kinetic energies to which deuterons and $\alpha$ particles can be accelerated? (b) Compute the mass electronic stopping power in water ($I_{\text{water}} = 78.0$ eV) for such proton assuming the shell, Bloch, and Barkas corrections to be negligible. Note that the Bragg rule needs to be used to determine $(Z/A)_{\text{water}}$. Compare with the $S_{el}/\rho$ value in the Data Tables.
*Answer: (a) E(deuteron) = 500.25 MeV; E($\alpha$) = 1007 MeV; (b) $S_{el}/\rho$(proton) = 2.20104 MeV cm$^2$ g$^{-1}$*

**Solution:**

(a) The approximate maximum kinetic energies are:

$$\text{Proton}: \quad \frac{z^2}{m_0c^2} = \frac{1^2}{938.272} = 1.0658 \times 10^{-3} \text{ MeV}$$

$$\text{Deuteron}: \quad \frac{z^2}{m_0c^2} = \frac{1^2}{1875.613} = 5.3316 \times 10^{-3}$$

$$\Rightarrow \frac{5.3316 \times 10^{-3}}{1.0658 \times 10^{-3}} \times 1000 \text{ MeV} = 500.25 \text{ MeV}$$

$$\text{Alpha}: \quad \frac{z^2}{m_0c^2} = \frac{2^2}{3727.379} = 1.0731 \times 10^{-3}$$

$$\Rightarrow \frac{1.0731 \times 10^{-3}}{1.0658 \times 10^{-3}} \times 1000 \text{ MeV} = 1006.9 \text{ MeV}$$

(b) A proton has a rest energy $m_0c^2 = 938.272$ MeV and charge $z = 1$. The Bragg rule for water yields $Z/A = 0.555087$, and we use the mean excitation energy $I_{\text{water}} = 78$ eV. The expression for the mass electronic stopping power is

$$S_{el}/\rho = 4\pi r_e^2 m_e c^2 N_A \times \frac{Z}{A}\frac{z^2}{\beta^2} \times L(\beta)$$

where the constant $4\pi r_e^2 m_e c^2 N_A = 0.307074$ MeV cm$^2$ g$^{-1}$.

$L(\beta)$ is the stopping number, $L(\beta) = L_0(\beta) + zL_1(\beta) + z^2L_2(\beta)$, and we assume $L_1(\beta)$ and $L_2(\beta)$ to be negligible. Thus, the only component of importance is $L_0(\beta)$

$$L_0(\beta) = \ln\left[\frac{2m_ec^2\beta^2}{(1-\beta^2)I}\right] - \beta^2 - \frac{C(\beta)}{Z} - \frac{\delta(\beta)}{2}$$

where we also assume $C(\beta)$ to be negligible (note that all the corrections mentioned are important at low energies), but the density effect needs to be calculated.

If the kinetic energy of the proton particle is 1000 MeV, $\tau = 1.06579$ and $\beta = 0.875026$. The Sternheimer parameter $X$ is

$$X = 0.5\log_{10}[\tau(\tau+2)] = 0.257107,$$

which is between $X_0$ (0.24) and $X_1 = 2.8004$, and therefore the density effect is given by

$$\delta(X) = 4.6052X + a(X_1 - X)^m + C = 0.0238$$

With this density effect correction and the $\beta$ and $I$ values $L_0(\beta) = 9.88701$ and $S_{el}/\rho = 2.20104\,\text{MeV cm}^2\,\text{g}^{-1}$. This can be compared with the more accurate value in the Data Tables, 2.2021 MeV cm$^2$ g$^{-1}$, showing that the assumptions made on negligible contributions are a good approximation at the proton energy considered.

**12** Compare the Rutherford differential cross section (DCS) for inelastic scattering with that from Møller for a 10 MeV electron and an energy transfer of 2 MeV. How large is the overall correction for spin, relativistic, and exchange effects?

*Answer: The overall correction increases the DCS by about 7.5%.*

**Solution:**
The two DCS are, respectively,

$$\frac{d\sigma_{\text{Ruth}}^{\text{inel}}}{dW} = 2\pi r_e^2 m_e c^2 \frac{1}{\beta^2}\frac{1}{W^2}$$

$$\frac{d\sigma_{\text{Møller}}^{\text{inel}}}{dW} = 2\pi r_e^2 m_e c^2 \frac{1}{\beta^2}\frac{1}{W^2}$$

$$\left[1 + \frac{W^2}{(E-W)^2} + \frac{\tau^2}{(\tau+1)^2}\left(\frac{W}{E}\right)^2 - \frac{2\tau+1}{(\tau+1)^2}\frac{W}{E-W}\right]$$

For a 10 MeV electron, $\tau = 19.5695$ and $\beta = 0.998818$. For $W = 2$ MeV, we get

$$\frac{d\sigma_{\text{Ruth}}^{\text{inel}}}{dW} = 6.38897 \times 10^{-26}\,\text{cm}^2\text{MeV}^{-1}$$

$$\frac{d\sigma_{\text{Møller}}^{\text{inel}}}{dW} = 6.86807 \times 10^{-26}\,\text{cm}^2\text{MeV}^{-1}$$

that is, their ratio is 1.07499, that is, a 7.5% difference that corresponds to the overall correction within the square brackets of $d\sigma_{\text{Møller}}^{\text{inel}}/dW$.

**13** Calculate the mass electronic stopping power for electrons and positrons with a kinetic energy of 50 MeV in aluminium. Verify the influence of the density effect correction.

*Answer:*

With $\delta \neq 0$: $S_{el}(electron) = 1.79177\,MeV\,cm^2\,g^{-1}$
$\qquad\qquad S_{el}(positron) = 1.74363\,MeV\,cm^2\,g^{-1}$
With $\delta = 0$: $S_{el}(electron) = 2.16416\,MeV\,cm^2\,g^{-1}$
$\qquad\qquad S_{el}(positron) = 2.11603\,MeV\,cm^2\,g^{-1}.$

**Solution:**

The mass electronic stopping power, common to both light particles, is given by (neglecting the shell correction)

$$\frac{1}{\rho}S_{el}^{\pm} = 2\pi r_e^2 m_e c^2 N_A \frac{Z}{A}\frac{1}{\beta^2}[\ln(E/I)^2 + \ln(1+\tau/2) + F^{\pm}(\tau) - \delta(\beta)]$$

with

$$F^{-}(\tau) = (1-\beta^2)\left[1 + \frac{\tau^2}{8} - (2\tau+1)\ln 2\right]$$

$$F^{+}(\tau) = 2\ln 2 - \frac{\beta^2}{12}\left[23 + \frac{14}{\tau+2} + \frac{10}{(\tau+2)^2} + \frac{4}{(\tau+2)^3}\right]$$

for electrons and positrons, respectively.

For the aluminium target, we have $Z = 13, A = 26.98, \rho = 2.6989$ g cm$^{-3}$, $I_{Al} = 166$ eV, and the Sternheimer density effect parameters $C = -4.2395$, $X_0 = 0.1708$, $X_1 = 3.0127$, $a = 0.08024$, $m = 3.6345$, and $\delta(X_0) = 0.12$.

For a 50 MeV incident electron or positron, $\tau = 97.8476$ and $\beta = 0.999949$. The density effect parameter $X$ is

$$X = 0.5\log_{10}[\tau(\tau+2)] = 1.9949,$$

which is between $X_0$ and $X_1$, and therefore the density effect is given by

$$\delta(X) = 4.6052X + a(X_1 - X)^m + C = 5.03316$$

With this density effect correction, $\beta$ and $I$, and $2\pi r_e^2 m_e c^2 N_A = 0.153537$ MeV cm$^2$, results $F^{-}(\tau) = 0.108632$ and $F^{+}(\tau) = -0.541943$, and therefore

$$S_{el}/\rho(electron) = 1.79177 \text{ MeV cm}^2 \text{ g}^{-1}$$

$$S_{el}/\rho(positron) = 1.74363 \text{ MeV cm}^2 \text{ g}^{-1}$$

Should the density effect correction be neglected, that is, $\delta = 0$, the corresponding stopping power values would be

$$S_{el}/\rho(electron, \delta = 0) = 2.16416 \text{ MeV cm}^2 \text{ g}^{-1}$$

$$S_{el}/\rho(positron, \delta = 0) = 2.11603 \text{ MeV cm}^2 \text{ g}^{-1}$$

that is, for the present case, the δ-correction decreases the mass electronic stopping power by about 17%.

As a verification, our result for the density effect correction can be compared with the more accurate value in the Data Tables, $\delta = 5.0679$. This would yield 1.7892 MeV cm$^2$ g$^{-1}$ and 1.74106 MeV cm$^2$ g$^{-1}$ for electrons and positrons, respectively.

**14**  Calculate the most probable energy of a 50 MeV electron after traversing an aluminium layer of 5 mm; use the Landau straggling constant $K = 0.373$. Compare with the estimated mean energy obtained from the stopping power calculated in the previous exercise.
*Answer: $E_p = 48.2122\,MeV$, $\bar{E} = 47.5821\,MeV$.*

**Solution:**
We assume the path length to be identical to the layer thickness. From above, we have that for a 50 MeV electron, $\tau = 97.8476$ and $\beta = 0.999949$; the density effect was estimated to be $\delta = 5.03316$ and $I_{Al} = 166$ eV.
The most probable energy loss for a thickness $t = 0.5$ cm $\times 2.6989$ g cm$^{-3}$ = $1.34945$ g cm$^{-2}$ is given by

$$\Delta E_p = \varsigma \left[ \ln \left( \frac{2m_e c^2 \beta^2 \varsigma}{(1-\beta^2)I^2} \right) - \beta^2 - K - \delta(\beta) \right]$$

where

$$\varsigma = 2\pi r_e^2 m_e c^2 N_A \frac{Z}{A} \frac{z^2}{\beta^2} \, t = 0.0998429 \text{ MeV}$$

$$K = 0.373,$$

yielding $\Delta E_p = 1.78775$ MeV. Therefore, $E_p = E_0 - \Delta E_p = 48.2122$ MeV.
The mean energy loss is given by the mass electronic stopping power, estimated in the previous exercise to be $S_{el}/\rho = 1.79177$ MeV cm$^2$ g$^{-1}$. For the thickness $t$, $\Delta E_{mean} = S_{el}/\rho \times t = 2.4179$ MeV. Therefore $\bar{E} = E_0 - \Delta E_{mean} = 47.5821$ MeV.
Thus the energy of the peak of the straggling distribution is larger than the mean energy of the distribution.

**15**  Calculate the restricted mass electronic stopping powers in aluminium for electrons and positrons with kinetic energy of 50 MeV for a cut-off $\Delta = 15$ keV.
*Answer:*
*$L/\rho(\Delta = 15\,keV, electron) = 1.21215\,MeV\,cm^2\,g^{-1}$,*
*$L/\rho(\Delta = 15\,keV, positron) = 1.21211\,MeV\,cm^2\,g^{-1}$.*

**Solution:**
The mass restricted electronic stopping power, common to both light particles, is given by (neglecting the shell correction)

$$\frac{1}{\rho}L(E, \Delta) = 2\pi r_e^2 m_e c^2 N_A \frac{Z}{A} \frac{1}{\beta^2} [\ln (E/I)^2 + \ln(1 + \tau/2) + G^{\pm}(\tau, \eta) - \delta(\beta)]$$

where $\eta$ is the ratio of $\Delta$ to the kinetic energy of the incident particle ($\eta = \Delta/E$) and

$$G^-(\tau, \eta) = -1 - \beta^2 + \ln[4(1 - \eta)\eta] + (1 - \eta)^{-1}$$
$$+ (1 - \beta^2)[\tau^2 \eta^2/2 + (2\tau + 1)\ln(1 - \eta)]$$
$$G^+(\tau, \eta) = \ln 4\eta - \beta^2[1 + (2 - \xi^2)\eta - (3 + \xi^2)(\xi\tau/2)\eta^2$$
$$+ (1 + \xi\tau)(\xi^2\tau^2/3)\eta^3 - (\xi^3\tau^3/4)\eta^4]$$

for electrons and positrons, respectively, with $\xi = (\tau + 2)^{-1}$.

For the projectile and target, the data are the same as in the previous exercise, and so is the density effect correction. Therefore

$$G^-(\tau, \eta) = -7.72534$$
$$G^+(\tau, \eta) = -7.72593$$

and

$$L/\rho(\Delta = 15 \text{ keV, electron}) = 1.21215 \text{ MeV cm}^2 \text{ g}^{-1}$$
$$L/\rho(\Delta = 15 \text{ keV, positron}) = 1.21211 \text{ MeV cm}^2 \text{ g}^{-1}$$

The respective ratios to the unrestricted stopping powers (from the previous exercise) are

$$L/\rho(\Delta = 15 \text{ keV, electron})/S_{el}/\rho = 0.67651$$
$$L/\rho(\Delta = 15 \text{ keV, positron})/S_{el}/\rho = 0.69516$$

**16** Estimate the approximate mass radiative stopping power for the electrons in the previous exercise. Compare with the value given in the Data Tables.
*Answer: $S_{rad}$ (50 MeV electrons, Al) = 1.8815 MeV cm²g⁻¹*

**Solution:**
For this exercise we use the simple approximation of Bethe–Ashkin, that is,

$$\frac{1}{\rho}S_{rad} = 4\alpha r_e^2 N_A \frac{Z^2}{A} E_{tot} \left( \ln \frac{E_{tot}}{m_e c^2} - \frac{1}{3} \right)$$

where $E_{tot} = E + m_e c^2 = 50.511$ MeV. It is straightforward to get

$$4\alpha r_e^2 N_A \frac{Z^2}{A} = 0.00874349 \text{ MeV cm}^2 \text{ g}^{-1}$$

$$\ln \frac{E_{tot}}{m_e c^2} - \frac{1}{3} = 4.26025$$

$$S_{rad}/\rho = 1.8815 \text{ MeV cm}^2 \text{ g}^{-1}$$

Compared with the corresponding value in the Data Tables, 1.7609 MeV cm²g⁻¹, the approximation overestimates $S_{rad}/\rho$ by about 7%. It gets worse for higher atomic numbers, showing the need for using tabulated data.

**17** Using the radiation yield from the Data Tables, how much energy (J) is emitted as bremsstrahlung by $10^{15}$ electrons entering a layer of lead at $E = 10$ MeV and exiting at $E = 7$ MeV?
*Answer: 221.3 J*

**Solution:**
Radiation yield for Pb from tables: $Y(10 \text{ MeV}) = 0.31616$, $Y(7 \text{ MeV}) = 0.25435$.
Photon energy emitted per electron: $E_{ph} = (10 \times 0.31616) - (7 \times 0.25435) = 1.38115$ MeV
Total energy = 1.38115 MeV/electron × $10^{15}$ electron × $1.6018 \times 10^{-13}$ J/MeV = 221.285 J.

**18** After traversing one radiation length in air, an electron of 1 GeV has lost (a) 0.368 GeV, (b) none, and (c) 0.632 GeV?
*Answer: (c) 0.632 GeV*

**Solution:**
By definition, $E = E_0\, e^{-t/X_0}$, where $X_0$ is the radiation length. Thus, when $t = X_0$, $E = E_0\, e^{-1} = 0.368$ GeV. The energy loss is $\Delta E = 1 - 0.368 = 0.632$ GeV; hence the answer is (c).

**19** The probability for an energy loss $W$ in the interval $dW$ of a charged particle with kinetic energy $E$ and velocity $v$ in a single collision is proportional to (a) $W\,dW/E$, (b) $E\,dW$, and (c) $dW/(vW)^2$?
*Answer: (c) $dW/(vW)^2$*

**Solution:**
Under the classical framework for the inelastic scattering of a particle with charge $ze$ and mass $m$, the energy transferred to an electron target, that is, with $Z = 1$, is given by (see Eq. (2.243))

$$W = \frac{q^2}{2m_e} = 2\left(\frac{e^2}{4\pi\varepsilon_0}\right)^2\left(\frac{z\,Z}{v\,b}\right)^2\frac{1}{m_e} = 2r_e^2 m_e c^2 \frac{z^2}{(v\,b)^2}\frac{1}{m_e}$$

where $b$ is the impact parameter. From here

$$dW = -4\,r_e^2 z^2 \frac{m_e c^2}{m_e}\frac{db}{v^2 b^3} = -C\,\frac{db}{v^2 b^3}$$

where $C = 4\,r_e^2 z^2 c^2$ is constant.

The probability of colliding with an electron with impact parameter in the interval $[b, b + db]$ is

$$d\sigma = 2\pi\,b\,|db| = \frac{2\pi v^2 b^4}{C}\,dW = \frac{\pi\,C\,dW}{2(vW)^2} \propto \frac{dW}{(vW)^2}$$

Therefore the answer is (c).

# 3

# Uncharged Particle Interactions with Matter

1   On the basis of the Klein–Nishina (KN) theory, what is the ratio of the Compton interaction cross sections per atom for lead and carbon?
*Answer: The ratio of their atomic numbers, that is 82/6, since all electrons are assumed to be unbound and to have the same interaction cross section.*

**Solution:**
Recall that $_e\sigma_{C,KN}$ is the Klein–Nishina cross section *per electron*, that is, it has units of $cm^2$ per electron, and that the cross section *per atom* is defined by

$$_a\sigma_{C,KN} = Z \,_e\sigma_{C,KN} \quad (cm^2/atom)$$

Hence, as $_e\sigma_{C,KN}$ is independent of the atomic number and therefore is the same for all elements or compounds, the ratio of the $_a\sigma_{C,KN}$ cross sections for two materials is simply given by the ratio of their atomic numbers, that is, 82/6.

2   On the basis of the KN theory, which Compton mass attenuation coefficient is larger, that for carbon or for lead? Why?
*Answer: For carbon, because of its larger number of electrons per gram $(N_A Z/A)$*

**Solution:**
The Compton KN mass attenuation coefficient is defined by

$$\frac{\mu_C}{\rho} = \frac{N_A}{A} \,_a\sigma_{C,KN} = \frac{N_A Z}{A} \,_e\sigma_{C,KN} \quad (cm^2 \, g^{-1})$$

where $N_e = N_A Z/A$ is the number of electrons per gram of material (recall that the electron density is given by $n_e = \rho \, N_e = \rho \, N_A Z/A$).
Hence, carbon has the highest $\mu_C/\rho$ value because of its larger $N_e$, that is, in $N_A Z/A$, $Z/A = 0.4995$ for carbon, but it is 0.3958 for lead.

3   Considering that in a $\mu/\rho$ plot for the different photon interaction coefficients Rayleigh scattering is quite significant, should it be included in $\mu_{tr}/\rho$ and $\mu_{en}/\rho$ plots? Why?
*Answer: No, because Rayleigh scattering transfers no energy to charged particles*

*Fundamentals of Ionizing Radiation Dosimetry: Solutions to Exercises,* First Edition.
Pedro Andreo, David T. Burns, Alan E. Nahum, and Jan Seuntjens.
© 2017 Wiley-VCH Verlag GmbH & Co. KGaA. Published 2017 by Wiley-VCH Verlag GmbH & Co. KGaA.

4   Calculate the energy of the Compton-scattered photon at $\theta = 0°, 45°, 90°$, and $180°$ for $k = 50$ keV, $500$ keV, and $5$ MeV.
*Answer: At*

$$\theta = 0° : k' = k$$
$$45° : k' = 0.0486, 0.389, 1.293 \ MeV$$
$$90° : k' = 0.0455, 0.253, 0.464 \ MeV$$
$$180° : k' = 0.0418, 0.169, 0.243 \ MeV$$

**Solution:**
Using Eqs. (3.65) and (3.66) for the Compton-scattered photon, that is,

$$k' = \frac{k}{1 + \frac{k}{m_e c^2}(1 - \cos\theta)}$$

with $m_e c^2 = 0.511$ MeV, yields directly the *Compton line* values above.

5   What are the corresponding energies and angles of the recoiling electrons for the cases in the previous exercise?
*Answer: At $\theta = 0°, E = 0$ for all $k$*
$\theta = 45°, 0.001 \ MeV (65.5°); 0.111 \ MeV (50.7°); 3.707 \ MeV (12.6°)$
$\theta = 90°, 0.004 \ MeV (42.3°); 0.247 \ MeV (26.8°); 4.536 \ MeV (5.30°)$
$\theta = 180°, 0.008 \ MeV (0°); 0.331 \ MeV (0°); 4.757 \ MeV (0°)$

**Solution:**
The Compton electron recoil kinetic energy is given by $E = k - k'$, Eq. (3.68), where the expression for the energy of the scattered photon, $k'$, is given in the previous exercise.
For the Compton electron recoil angle, $\phi$, its relation with the photon-scattered angle, $\theta$, is given by Eq. (3.69):

$$\cot\phi = \frac{1}{\tan\phi} = \left(1 + \frac{k}{m_e c^2}\right)\tan\frac{\theta}{2}$$

Hence

| $k$ (MeV) | 0.05 MeV | 0.5 MeV | 5 MeV |
|---|---|---|---|
| $\theta°$ | Electron kinetic energy, $E$ (MeV), and angle, $\phi(°)$ | | |
| 0° | 0 | 0 | 0 |
| 45° | $1.393 \times 10^{-3}$ MeV (65.5°) | 0.111 MeV (50.7°) | 3.707 MeV (12.6°) |
| 90° | $4.456 \times 10^{-3}$ MeV (42.3°) | 0.247 MeV (26.8°) | 4.536 MeV (5.30°) |
| 180° | $8.183 \times 10^{-3}$ MeV (0°) | 0.331 MeV (0°) | 4.757 MeV (0°) |

**6** Calculate the total KN cross section from Eq. (3.84) for 1 MeV photons, and derive the Compton mass attenuation coefficient for copper, expressed in cm$^2$/g$^{-1}$ and m$^2$/kg$^{-1}$.
*Answer:* $_e\sigma_{C,KN} = 2.112 \times 10^{-25}$ *cm$^2$ per electron;*
$(\sigma_{C,KN}/\rho)_{Cu} = 0.0580$ *cm$^2$ g$^{-1}$ = 0.00580 m$^2$ kg$^{-1}$.*

**Solution:**
Equation (3.84) has the form

$$_e\sigma_{C,KN}$$

$$= 2\pi \int_{\theta=0}^{\pi} \frac{d_e\sigma_{C,KN}}{d\Omega_\theta} \sin\theta \; d\theta$$

$$= \pi r_e^2 \int_0^\pi \left(\frac{k'}{k}\right)^2 \left(\frac{k'}{k} + \frac{k}{k'} - \sin^2\theta\right) \sin\theta \; d\theta$$

$$= 2\pi r_e^2 \left\{ \frac{1+\epsilon}{\epsilon^2} \left[ \frac{2(1+\epsilon)}{1+2\epsilon} - \frac{\ln(1+2\epsilon)}{\epsilon} \right] + \frac{\ln(1+2\epsilon)}{2\epsilon} - \frac{1+3\epsilon}{(1+2\epsilon)^2} \right\}$$

The photon energy in terms of m$_e$c$^2$ is

$$\epsilon = \frac{1 \text{ MeV}}{0.511 \text{ MeV}} = 1.9569$$

and the Klein–Nishina cross section then becomes

$$_e\sigma_{C,KN} = 2\pi(2.818 \times 10^{-13})^2 \left\{ \frac{2.9569}{1.9569^2} \left[ \frac{2 \times 2.9569}{4.9138} - \frac{\ln 4.9138}{1.9569} \right] \right.$$

$$\left. + \frac{\ln 4.9138}{3.9138} - \frac{6.8707}{4.9138^2} \right\}$$

$$= 2.112 \times 10^{-25} \text{ cm}^2 \text{ per electron,}$$

which coincides with the value in electronic Table A.8 of the Data Tables. For copper ($Z = 29, A = 63.546$), we have that

$$\frac{Z N_A}{A} = \frac{29 \times 6.022 \times 10^{23}}{63.546} = 2.748 \times 10^{23} \frac{\text{electron}}{\text{g}}$$

Therefore, the mass attenuation coefficient is given by

$$\frac{\sigma_{C,KN}}{\rho} = \frac{Z N_A}{A} {}_e\sigma_{C,KN} = 2.112 \times 10^{-25} \frac{\text{cm}^2}{\text{electron}}$$

$$\times 2.748 \times 10^{23} \frac{\text{electron}}{\text{g}}$$

$$= 0.05804 \frac{\text{cm}^2}{\text{g}} = 0.005804 \frac{\text{m}^2}{\text{kg}}$$

**7** Repeat the previous calculation of $\sigma_{C,KN}/\rho$ for 1 MeV photons in carbon and lead. Compare the values obtained for C, Cu, and Pb with those derived from the Compton cross sections in the electronic Data Tables and identify the reasons for the discrepancy.

*Answer: $\sigma_{C,KN}/\rho = 0.6353$ for C, $0.05804$ for Cu, and $0.05033$ cm²/g for Pb, whereas $\sigma_C/\rho = 0.6352$, $0.05793$, and $0.04987$, respectively. Hence the ratios electronic tables/KN are 1.000, 0.998, and 0.991 for C, Cu, and Pb, respectively. The reason for the discrepancy is that the tabulated data include binding effects and Doppler broadening, not included in KN, which increase with Z (especially when the energy decreases).*

**Solution:**
From the previous exercise, the Klein–Nishina cross section for 1 MeV photons is

$$_e\sigma_{C,KN} = 2.112 \times 10^{-25} \frac{cm^2}{electron}$$

and, as above, the Compton mass attenuation coefficient is given by

$$\frac{\sigma_{C,KN}}{\rho} = \frac{Z N_A}{A} {}_e\sigma_{C,KN}$$

The values of $\sigma_{C,KN}/\rho$ for each element become

| Element | Z | A | $ZN_A/A \times 10^{23}$ | $\sigma_{C,KN}/\rho$ cm²/g⁻¹ |
|---------|---|---|--------------------------|-------------------------------|
| Carbon | 6 | 12.011 | 3.008 | 0.06353 |
| Copper | 29 | 63.546 | 2.748 | 0.05804 |
| Lead | 82 | 207.20 | 2.383 | 0.05033 |

Proceeding in the same way, but now using the Compton cross sections in the Data Tables, which correspond to $_a\sigma_C$ values, that is, $Z\, {}_e\sigma_C$, in square centimeter per atom, and then need to be multiplied only by $N_A/A$

| Element | $_a\sigma_C$ Data Tables | $\sigma_C/\rho$ cm² g⁻¹ | Ratio to $\sigma_{C,KN}/\rho$ |
|---------|---------------------------|--------------------------|-------------------------------|
| Carbon | $1.2669 \times 10^{-24}$ | 0.06352 | 1.000 |
| Copper | $6.1131 \times 10^{-24}$ | 0.05793 | 0.998 |
| Lead | $1.7160 \times 10^{-23}$ | 0.04987 | 0.991 |

As the $_a\sigma_C$ tabulated data include binding effects and Doppler broadening (together they form the impulse approximation), which are not included in the KN cross sections, the ratios provide an estimate of the combination of these two corrections. At 1 MeV, for low $Z$ elements, the corrections are entirely negligible and become $\sim 1\%$ for Pb.

Note that, as shown in Figure 3.24 for aluminium, the difference in the two types of cross sections is substantial below about 100 keV.

**8** What is (a) the maximum energy and (b) the average energy of the Compton recoil electrons generated by 20 keV and 20 MeV $\gamma$ rays?

*Answer:*

$$20 \ keV : E_{max} = 1.45 \ keV, \qquad \bar{E} = 0.721 \ keV$$
$$20 \ MeV : E_{max} = 19.75 \ MeV, \quad \bar{E} = 14.5 \ MeV$$

**Solution:**

(a) The maximum energy transfer to the Compton recoil electron for a given photon energy $k$ occurs at $\theta = \pi$ (photon backscattering), which corresponds to an electron recoil angle $\phi = 0$. It is obtained from Eq. (3.68) that

$$E_{max} = \left[ \frac{k \, \epsilon (1 - \cos\theta)}{1 + \epsilon(1 - \cos\theta)} \right]_{\theta=\pi} = \frac{2k^2}{2k + 0.511 \ \text{MeV}}$$

where $\epsilon = k/m_e c^2$. Substituting values results

$$\text{for} \quad k = 0.02 \ \text{MeV} \quad E_{max} = 1.45 \times 10^{-3} \ \text{MeV}$$
$$k = 20 \ \text{MeV} \quad E_{max} = 19.75 \ \text{MeV}$$

(b) The mean energy transfer for the Compton effect is given by

$$\bar{E} = k \frac{{}_e\sigma^{tr}_{C,KN}}{{}_e\sigma_{C,KN}}$$

From the KN Table A.8 in the Data Tables,

| At | ${}_e\sigma^{tr}$ | ${}_e\sigma$ |
|---|---|---|
| 20 keV | $0.2228 \times 10^{-25}$ | $0.6180 \times 10^{-24}$ |
| 20 MeV | $0.2198 \times 10^{-25}$ | $0.3025 \times 10^{-25}$ |

Hence

$$\bar{E} = 0.721 \ \text{keV} \ \text{ and } \ 14.5 \ \text{MeV}.$$

**9** (a) Calculate the energy of a photoelectron ejected from the K-shell in rhodium by a 50 keV photon. (b) Calculate the energy-transfer coefficient $\mu_{ph,tr}/\rho = \bar{f}_{ph} \ \mu_{ph}/\rho$.
*Answer: $E = 26.78 \ keV$; $\mu_{ph,tr}/\rho = 5.70 \ cm^2 \ g^{-1}$*

**Solution:**

(a) A photon with energy $k$ interacts with a tightly bound orbital electron. The photon is absorbed completely and the electron is ejected as a photoelectron with a kinetic energy $E = k - U_B$, where $U_B$ is the binding energy of the atomic shell from which the electron is ejected.

From Table A.5 in the Data Tables, for rhodium $(U_B)_K = 23.225$ keV; therefore the electron energy is

$$E = 50 - 23.225 = 26.775 \ \text{keV}$$

(b) The photoelectric mass energy-transfer coefficient is given by (see Eqs. (3.159)–(3.161))

$$\frac{\mu_{ph,tr}}{\rho} = \frac{\mu_{ph}}{\rho}\,\bar{f}_{ph} = \frac{\mu_{ph}}{\rho}\left[\frac{k - \bar{k}_{ph}(k)}{k}\right] = \frac{\mu_{ph}}{\rho}\left[1 - \frac{\sum_j p_j(k)\,\omega_j\,\bar{k}_j}{k}\right]$$

where $\bar{k}_{ph}(k) = \sum p_j(k)\,\omega_j\,\bar{k}_j$ represents the average photon energy emitted as fluorescence after a photoelectric interaction. The summation is over all shells $j$ involved in the photoelectric process. The quantities $p_j(k)$, $\omega_j$, and $\bar{k}_j$ are, respectively, the probability for the photoeffect to take place in shell $j$, the fluorescent yield, and the average photon energy emitted.

From Table A.5 the values of $p_j(k)$, $\omega_j$, and $\bar{k}_j$ for rhodium ($Z = 45, A = 102.906$) are (assuming that only the shells K, $L_1$, and $M_1$ participate in the event)

| Shell | $p_j(k)$ | $\omega_j$ | $\bar{k}_j$ (keV) | Product (keV) |
|---|---|---|---|---|
| K | 0.85505 | $8.808\ 10^{-1}$ | 20.738 | 14.33 |
| L | 0.82337 | $5.07\ 10^{-2}$ | 2.936 | 0.12 |
| M | 0.91863 | $5.044\ 10^{-4}$ | 0.458 | $2.29\ 10^{-4}$ |
| $\bar{k}_{ph}(k)$ | | | Sum= | 14.45 |

From the photon cross sections in the Data Tables

$$_a\mu_{ph} = 1.36965 \times 10^{-21}\ \text{cm}^2\ \text{per atom}$$

and multiplying by $N_A/A = 5.852 \times 10^{21}$, the mass attenuation coefficient is

$$\frac{\mu_{ph}}{\rho} = {_a\mu_{ph}}\,\frac{N_A}{A} = 8.015\ \text{cm}^2\ \text{g}^{-1}$$

Therefore the mass-energy transfer coefficient is

$$\frac{\mu_{ph,tr}}{\rho} = 8.015\left[\frac{50 - 14.45}{50}\right] = 5.699\ \text{cm}^2\ \text{g}^{-1}.$$

**10** What is the average energy of the charged particles resulting from pair production in (a) the nuclear field and (b) the electron field for photons of $k = 2$ MeV and 20 MeV?

*Answer: At*

$$k = 2\ MeV: \quad \bar{E} = 0.489 \qquad 0\ MeV$$
$$k = 20\ MeV: \quad \bar{E} = 9.49 \qquad 6.33\ MeV$$

**Solution:**

(a) For a threshold of $2\ m_e c^2 = 1.022$ MeV, the energy transfer to charged particles in pair production generally is not equal for the electron and the positron, but their average kinetic energy is given by Eq. (3.123):

$$\bar{E}_{pp} = \frac{k - 2\,m_e c^2}{2}$$

Therefore

| k | $\bar{E}_{pp}$ (MeV) |
|---|---|
| 2 MeV | 0.489 |
| 20 MeV | 9.49 |

(b) Similarly, for triplet production, the average kinetic energy is given by Eq. (3.127):

$$\bar{E}_{tp} = \frac{k - 2\,m_e c^2}{3}$$

Hence

| k | $\bar{E}_{tp}$ (MeV) |
|---|---|
| 2 MeV | Below threshold |
| 20 MeV | 6.33 |

**11** A narrow beam containing $10^{20}$ photons at 6 MeV impinges perpendicularly on a layer of lead 12 mm thick. How many interactions of each type (Rayleigh, Compton, photoelectric, pair, triplet) occur in the lead?
*Answer: Rayleigh: $7.21 \times 10^{16}$; Compton: $1.789 \times 10^{19}$; photoelectric: $9.764 \times 10^{17}$; pair: $2.586 \times 10^{19}$; triplet: $1.219 \times 10^{17}$*

**Solution:**
For Pb, $Z = 82, A = 207.2$, and $\rho = 11.35$ g cm$^{-3}$. Hence $N_A/A = 2.906 \times 10^{21}$. From the photon cross section shown in Data Tables for Pb, the partial atomic cross sections at 6 MeV are

| Interaction | $_a\mu_j$ (cm$^2$ per atom) |
|---|---|
| Rayleigh | $2.5896 \times 10^{-26}$ |
| Compton | $6.0018 \times 10^{-24}$ |
| Photoelectric | $3.2749 \times 10^{-25}$ |
| Pair prod | $8.6727 \times 10^{-24}$ |
| Triplet prod | $4.0875 \times 10^{-26}$ |
| Total $_a\mu = \sum_j \,_a\mu_j =$ | $1.5069 \times 10^{-23}$ |

Thus,

$$\mu/\rho = {}_a\mu \frac{N_A}{A} = 1.5069 \times 10^{-23} \times 2.906 \times 10^{21} = 4.38 \times 10^{-2} \text{ cm}^2 \text{ g}^{-1}$$

The number of photons interacting is

$$\Delta N = N_0 - N = 10^{20}(1 - e^{-4.38\times10^{-2}\times11.35\times1.2})$$
$$= 10^{20}(1 - 0.5507) = 4.493 \times 10^{19}$$

Therefore the numbers of interactions of each type will be

$$\text{Rayleigh}: \ 4.493 \times 10^{19} \times \left( \frac{2.5896 \times 10^{-26}}{1.5069 \times 10^{-23}} \right) = 7.721 \times 10^{16}$$

$$\text{Compton}: \ 4.493 \times 10^{19} \times \left( \frac{6.0018 \times 10^{-24}}{1.5069 \times 10^{-23}} \right) = 1.789 \times 10^{19}$$

$$\text{Photoelectric}: \ 4.493 \times 10^{19} \times \left( \frac{3.2749 \times 10^{-25}}{1.5069 \times 10^{-23}} \right) = 9.764 \times 10^{17}$$

$$\text{Pair}: \ 4.493 \times 10^{19} \times \left( \frac{8.6727 \times 10^{-24}}{1.5069 \times 10^{-23}} \right) = 2.586 \times 10^{19}$$

$$\text{Triplet}: \ 4.493 \times 10^{19} \times \left( \frac{4.0875 \times 10^{-26}}{1.5069 \times 10^{-23}} \right) = 1.219 \times 10^{17}$$

**12** Assuming that each interaction in the previous exercise results in one primary photon being removed from the beam, how much energy (in Joule) is removed by each type of interaction?
*Answer: Rayleigh: 7.421 × 10⁴; Compton: 1.720 × 10⁷; photoelectric: 9.385 × 10⁵; pair: 2.485 × 10⁷; triplet: 1.171 × 10⁵*

**Solution:**

$$\text{Rayleigh}: \ 7.721 \times 10^{16} \ \text{phot} \times 6\frac{\text{MeV}}{\text{phot}}$$
$$\times 1.602 \times 10^{-13} \ \frac{\text{J}}{\text{MeV}} = 7.421 \times 10^4 \ \text{J}$$

$$\text{Compton}: \ 1.789 \times 10^{19} \ \text{phot} \times 6\frac{\text{MeV}}{\text{phot}}$$
$$\times 1.602 \times 10^{-13} \ \frac{\text{J}}{\text{MeV}} = 1.720 \times 10^7 \ \text{J}$$

$$\text{Photoelectric}: \ 9.764 \times 10^{17} \ \text{phot} \times 6\frac{\text{MeV}}{\text{phot}}$$
$$\times 1.602 \times 10^{-13} \ \frac{\text{J}}{\text{MeV}} = 9.385 \times 10^5 \ \text{J}$$

$$\text{Pair}: \ 2.586 \times 10^{19} \ \text{phot} \times 6\frac{\text{MeV}}{\text{phot}}$$
$$\times 1.602 \times 10^{-13} \ \frac{\text{J}}{\text{MeV}} = 2.485 \times 10^7 \ \text{J}$$

$$\text{Triplet}: \ 1.219 \times 10^{17} \ \text{phot} \times 6\frac{\text{MeV}}{\text{phot}}$$
$$\times 1.602 \times 10^{-13} \ \frac{\text{J}}{\text{MeV}} = 1.171 \times 10^5 \ \text{J}$$

**13** How much energy is transferred to charged particles by each type of interaction in the above two exercises?

*Answer: Rayleigh: 0; Compton: $1.108 \times 10^7$ J; photoelectric: $9.281 \times 10^5$ J;
pair: $2.062 \times 10^7$ J; triplet: $\sim 9.718 \times 10^4$ J*

**Solution:**
(a) Rayleigh: No energy transfer to electrons
(b) Compton:

$$\frac{\bar{E}}{k} \times 1.720 \times 10^7 \text{ J} = \frac{\mu_{C,tr}}{\mu_C} \times 1.720 \times 10^7 \text{ J}$$

$$= \frac{4.716 \times 10^{-26}}{7.323 \times 10^{-26}} \times 1.720 \times 10^7 \text{ J} = 1.108 \times 10^7 \text{ J}$$

(c) Photoelectric:

The photoeffect energy-transfer fraction is given by (see Eq. (3.161))

$$\bar{f}_{ph} = \frac{k - \bar{k}_{ph}(k)}{k} = 1 - \frac{\sum_j p_j(k)\,\omega_j\,\bar{k}_j}{k}$$

where $\bar{k}_{ph}(k) = \sum p_j(k)\,\omega_j\,\bar{k}_j$ represents the average photon energy emitted as fluorescence after a photoelectric interaction. The summation
is over all shells $j$ involved in the photoelectric process. The quantities
$p_j(k)$, $\omega_j$, and $\bar{k}_j$ are, respectively, the probability for the photoeffect to
take place in shell $j$, the fluorescent yield, and the average photon energy
emitted.

From Table A.5 the values of $p_j(k)$, $\omega_j$, and $\bar{k}_j$ for lead are (assuming that
only the shells K, $L_1$, and $M_1$ participate in the event)

| Shell | $p_j(k)$ | $\omega_j$ | $\bar{k}_j$ (keV) | Product (keV) |
|---|---|---|---|---|
| K | 0.80367 | $9.61 \ 10^{-1}$ | 81.248 | 62.75 |
| L | 0.75546 | $3.86 \ 10^{-1}$ | 12.913 | 3.77 |
| M | 0.79301 | $3.55 \ 10^{-2}$ | 2.921 | $8.22 \ 10^{-2}$ |
| $\bar{k}_{ph}(k)$ | | | Sum = | 66.60 |

Hence

$$\left(\frac{k - \bar{k}_{ph}(k)}{k}\right) \times 9.385 \times 10^5 \text{ J}$$

$$= \left(\frac{6 - 66.60 \times 10^{-3}}{6}\right) \times 9.385 \times 10^5 \text{ J} = 9.281 \times 10^5 \text{ J}$$

(d) Pair production (see Eq. (3.168)):

$$\left(\frac{k - 1.022}{k}\right) \times 2.485 \times 10^7 \text{ J} = 2.062 \times 10^7 \text{ J}$$

(e) Triplet production (see Eq. (3.169)), that is, the energy-transfer fraction is approximately the same as for pair production (but the energy transferred is different):

$$\sim \left( \frac{k - 1.022}{k} \right) \times 1.171 \times 10^5 \text{ J} = 9.718 \times 10^4 \text{ J}$$

**14** Determine the maximum and average energy transfer to the recoil nucleus in a single collision of a 2 MeV neutron with (a) deuterium and (b) lead. *Answer: (a) 1.78 MeV and 0.89 MeV; (b) 0.04 MeV and 0.02 MeV*

**Solution:**
Equations (3.181) and (3.182) give, respectively, the maximum and average energy transfer to the nucleus according to

$$Q_{max} = E_{n,i} \frac{4 \, m_n \, M}{(m_n + M)^2}$$

$$\bar{Q} = E_{n,i} \frac{2 \, m_n \, M}{(m_n + M)^2} = \frac{Q_{max}}{2}$$

where $E_{n,i}$ is the incident neutron energy and $m_n$ and $M$ are the neutron and recoil nucleus masses, respectively. As the masses enter as a ratio in the expressions, they can be replaced by their nucleon number, $A$.
Therefore, $m_n = 1$.
(a) For deuterium ($^2_1$H), $M = 2$

$$Q_{max} = 2 \text{ MeV} \frac{4 \times 1 \times 2}{(1 + 2)^2} = 1.778 \text{ MeV}$$

$$\bar{Q} = 2 \text{ MeV} \frac{2 \times 1 \times 2}{(1 + 2)^2} = 0.889 \text{ MeV}$$

that is, the neutron can transfer up to ~90% of its energy in a single collision with deuterium, as their masses are quite similar.
(b) For lead ($^{207}_{82}$Pb), $M = 207$

$$Q_{max} = 2 \text{ MeV} \frac{4 \times 1 \times 207}{(1 + 207)^2} = 0.038 \text{ MeV}$$

$$\bar{Q} = 2 \text{ MeV} \frac{2 \times 1 \times 207}{(1 + 207)^2} = 0.019 \text{ MeV}$$

that is, the neutron can transfer only up to ~2% of its energy in a single collision with lead, as their masses are substantially different.

**15** Derive expressions to estimate the average and minimum number of elastic interactions to decrease the energy of a neutron from an initial energy $E_{in}$ to a final energy $E_{fin}$.
*Answer:* $n_{av} = \dfrac{ln \frac{E_{fin}}{E_{in}}}{ln \left[ \frac{M^2 + m_n^2}{(M + m_n)^2} \right]}$; $n_{min} = \dfrac{ln \frac{E_{min,fin}}{E_{in}}}{ln \left[ \frac{M^2 - m_n^2}{(M + m_n)^2} \right]}.$

**Solution:**

The average neutron kinetic energy after one elastic collision is given by Eq. (3.183):

$$\bar{E}_f = E_i - \bar{Q} = E_i \left[ \frac{M^2 + m_n^2}{(M + m_n)^2} \right] \sim E_i \, [m - \text{ratio}]$$

which for successive interactions with energies $E_0, E_1, \cdots, E_n$ becomes

$$E_1 = E_0 \, [m - \text{ratio}]$$
$$E_2 = E_1 \, [m - \text{ratio}] = E_0 \, [m - \text{ratio}]^2$$
$$E_3 = E_2 \, [m - \text{ratio}] = E_0 \, [m - \text{ratio}]^3$$
$$\cdots$$
$$E_n = E_{n-1} \, [m - \text{ratio}] = E_0 \, [m - \text{ratio}]^n$$

that is, after $n_{av}$ collisions the average neutron kinetic energy is given by the expression

$$\bar{E}_{\text{fin}} = E_{\text{in}} \left[ \frac{M^2 + m_n^2}{(M + m_n)^2} \right]^{n_{av}}$$

Solving for $n_{av}$,

$$n_{av} = \frac{\ln \frac{\bar{E}_{\text{fin}}}{E_{\text{in}}}}{\ln \left[ \frac{M^2 + m_n^2}{(M + m_n)^2} \right]}$$

The minimum number of collisions will occur when the energy transferred to a nucleus in each collision is the maximum allowed, yielding the minimum neutron energy in each interaction. The minimum neutron kinetic energy can be written as

$$E_{\text{min},f} = E_i - Q_{\text{max}} = E_i \left[ \frac{M^2 - m_n^2}{(M + m_n)^2} \right] \sim E_i \, [m' - \text{ratio}]$$

and proceeding as above,

$$E_{\text{min, fin}} = E_{\text{in}} \left[ \frac{M^2 - m_n^2}{(M + m_n)^2} \right]^{n_{\text{min}}}$$

from where

$$n_{\text{min}} = \frac{\ln \frac{E_{\text{min, fin}}}{E_{\text{in}}}}{\ln \left[ \frac{M - M_n}{M + m_n} \right]}$$

**16** Estimate the average number of collisions that a 4 MeV neutron impinging on a target of (a) gold and (b) beryllium will experience to decrease its energy to 1 eV.

*Answer: (a) 1505; (b) 77*

**Solution:**
Using the expression derived in the previous exercise,

$$n_{av} = \frac{\ln \frac{\bar{E}_{fin}}{E_{in}}}{\ln \left[ \frac{M^2 + m_n^2}{(M + m_n)^2} \right]}$$

As above, with $m_n = 1$,
a) For gold ($^{197}_{79}$Au), $M = 197$:

$$n_{av} = \frac{\ln \frac{1 \times 10^{-6}}{4}}{\ln \left[ \frac{197^2 + 1}{(197 + 1)^2} \right]} = \frac{-15.202}{-0.010} = 1505$$

b) For beryllium ($^9_4$Be), $M = 9$:

$$n_{av} = \frac{\ln \frac{1 \times 10^{-6}}{4}}{\ln \left[ \frac{9^2 + 1}{(9 + 1)^2} \right]} = \frac{-15.202}{-0.1198} = 76.6$$

**17**  (a) Estimate the minimum number of collisions that a reactor 2 MeV neutron will suffer in a carbon moderator to become thermalized. (b) Estimate the same for a hydrogen moderator.
*Answer: (a) 55; (b) 1*

**Solution:**
The minimum number of collisions is given by the expression derived in a previous exercise:

$$n_{min} = \frac{\ln \frac{E_{fin}}{E_{in}}}{\ln \left[ \frac{M^2 - m_n^2}{(M + m_n)^2} \right]}$$

where for a thermal neutron $E_{fin} = 0.025$ eV.
(a) For carbon ($^{12}_6$C), $M = 12$:

$$n_{min} = \frac{\ln \frac{0.025 \times 10^{-6}}{2}}{\ln \left[ \frac{12^2 + 1}{(12 + 1)^2} \right]} = \frac{-18.20}{-0.334} = 54.5$$

(b) For hydrogen, as $m_n \approx M$, the neutron can lose all its energy in a single interaction, leaving the neutron at rest and the nucleus being ejected forward with the same energy that the incident neutron had.

# 4

# Field and Dosimetric Quantities and Radiation Equilibrium: Definitions and Interrelations

**1**  A broad plane-parallel beam of electrons is perpendicularly incident upon a thin foil that scatters the electrons through an average angle of 20°, stopping none of them. (a) What is the ratio of the fluence of primary electrons just behind the foil to that with the foil removed? (b) What is the ratio of the number of electrons per cm$^2$ passing through a plane just behind (and parallel to) the foil to that with the foil removed?
*Answer: (a) 1.06; (b) Unity*

**Solution:**

(a) See Figure 4.1 below (a reproduction of Figure 4.3 in the textbook). Scattering brings individual electron paths (assumed straight) closer together by the factor cos 20° = 0.94. Thus the number striking a given sphere behind the foil is increased by the factor 1/ cos 20° = 1.06.

(b) The spacing of electron tracks is the same in all planes that are parallel to the scattering foil.

**2**  The fluence rate decreases with increasing distance from a point source of γ rays as the inverse square of the distance (see Section 5.7 in the textbook). The strength of the electric field surrounding a point electric charge does likewise. At a point midway between two identical charges, the electric field is zero. (a) What is the fluence rate midway between two identical sources? (b) What is the essential difference between the two cases?
*Answer: (a) Twice that at the midway point due to one of the sources. (b) Fluence rate is a scalar quantity; electric field strength is a vector. Vector addition depends on orientation; scalar addition does not.*

**3**  A point source of $^{60}$Co γ rays emits equal numbers of photons of 1.17 MeV and 1.33 MeV, giving a fluence rate of $5.7 \times 10^9$ photons cm$^{-2}$ s$^{-1}$ at a specified location. What is the energy fluence rate there, expressed (a) in erg cm$^{-2}$ s$^{-1}$ and (b) in J m$^{-2}$ min$^{-1}$?
*Answer: (a) $1.14 \times 10^4$ erg cm$^{-2}$s$^{-1}$; (b) 685 J m$^{-2}$ min$^{-1}$.*

*Fundamentals of Ionizing Radiation Dosimetry: Solutions to Exercises,* First Edition.
Pedro Andreo, David T. Burns, Alan E. Nahum, and Jan Seuntjens.
© 2017 Wiley-VCH Verlag GmbH & Co. KGaA. Published 2017 by Wiley-VCH Verlag GmbH & Co. KGaA.

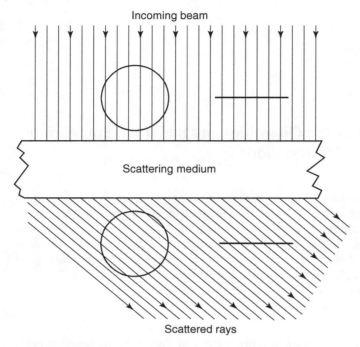

**Figure 4.1** Electrons scattered through an angle $\theta$ in a non-absorbing foil.

**Solution:**
For the two energies, $\dot{\Phi}_{tot} = 5.7 \times 10^9$ photons cm$^{-2}$ s$^{-1}$. Therefore,
$\dot{\Phi}_{1.17 \text{ or } 1.33} = \frac{5.7 \times 10^9}{2} = 2.85 \times 10^9$ photons cm$^{-2}$ s$^{-1}$
The energy fluence rate is

$$\dot{\Psi} = 2.85 \times 10^9 (1.17 + 1.33)$$

$$= 7.125 \times 10^9 \text{ photons cm}^{-2} \text{ s}^{-1} \text{ MeV per photon}$$

$$= 7.125 \times 10^9 \text{ MeV cm}^{-2} \text{ s}^{-1}$$

(a) $\dot{\Psi} \times 1.602 \times 10^{-6}$ erg MeV$^{-1}$ = $1.141 \times 10^4$ erg cm$^{-2}$ s$^{-1}$
(b) $\dot{\Psi} \times$ 1 J/$10^7$ erg $\times 10^4$ cm$^2$ /1 m$^2 \times$ 60 s/1 min = 685 J m$^{-2}$ min$^{-1}$

**4** In the previous exercise, what is the energy fluence of 1.17 MeV photons
during 24 h in (a) erg cm$^{-2}$ and (b) J m$^{-2}$.
*Answer: (a) 4.62 × 10$^8$ erg cm$^{-2}$; (b) 4.62 × 10$^5$ J m$^{-2}$*

**Solution:**
$\Psi = 5.7 \times 10^9 / 2$ photons cm$^{-2}$ s$^{-1} \times 1.17$ MeV/photon $\times$
$1.602 \times 10^{-6}$ erg/MeV = 5342 erg cm$^{-2}$ s$^{-1}$
(a) $\Psi \times 24$ h $\times$ 3600 s/1 h = $4.62 \times 10^8$ erg cm$^{-2}$
(b) Now taking $\Psi$ from (a), $\Psi \times$ 1 J/$10^7$ erg $\times 10^4$ cm$^2$/1 m$^2$ = $4.62 \times 10^5$ J m$^{-2}$

**5** As discussed in the textbook, the unit of energy fluence differential in energy, $\Psi_E$, coincides with that of fluence, $\Phi$, which sometimes leads to confusion. Show that the integral of the two quantities differential in energy yields the correct units of $\Phi$ and $\Psi$ and that absorbed dose calculations with photons (when $D \approx K_{el}$) and electrons have units of gray.

**Solution:**
The quantity fluence, defined in terms of number of particles per unit cross-sectional area, has the units

$$\Phi = \frac{1}{cm^2}$$

Energy fluence, related to the quantity fluence by $\Psi = E\,\Phi$, therefore has units

$$\Psi = \frac{eV}{cm^2}$$

The units of the two quantities differential in energy, or quantity divided by energy, become

$$\Phi_E = \frac{1}{cm^2\ eV}$$

$$\Psi_E = \frac{eV}{cm^2}\frac{1}{eV} = \frac{1}{cm^2}$$

that is, the unit of energy fluence differential in energy, $\Psi_E$, coincides with that of fluence, $\Phi$ (not with $\Phi_E$).
The integral of the two quantities differential in energy becomes

$$\Phi = \int \Phi_E\ dE \approx \sum_i \Phi_{E,i}\ \Delta E_i = \frac{1}{cm^2\ eV}\ eV = \frac{1}{cm^2}$$

$$\Psi = \int \Psi_E\ dE \approx \sum_i \Psi_{E,i}\ \Delta E_i = \frac{1}{cm^2}\ eV = \frac{eV}{cm^2}$$

that is, the units are identical to those given above.
For absorbed dose calculations with photons (when $D \approx K_{el}$) and electrons,

$$D_{\gamma,med} = \int \Psi_{E,med}(\mu_{en}/\rho)_{med}\ dE = \frac{1}{cm^2} \times \frac{cm^2}{g} \times eV = \frac{eV}{g}$$

$$= \int E\,\Phi_{E,med}(\mu_{en}/\rho)_{med}\ dE = eV \times \frac{1}{cm^2\ eV} \times \frac{cm^2}{g} \times eV = \frac{eV}{g}$$

$$D_{e,med} = \int \Phi_{E,med}(S_{el}/\rho)_{med}\ dE = \frac{1}{cm^2\ eV} \times \frac{eV}{g\ cm^{-2}} \times eV = \frac{eV}{g}$$

and eV/g can be converted to gray (J kg$^{-1}$) by multiplying by $1.6022 \times 10^{-19}$ J/1 eV and by 1000 g/1 kg.

**6** A certain Monte Carlo code (e.g., Penelope; see Chapter 8) outputs a tally 'FluenceTrackLength' for the spectral fluence per incident particle as tracks integrated over a given region volume in units of cm eV$^{-1}$. Explain how the user should derive the correct units of $\Phi_E$.

*Answer: Dividing by the volume of the region yields units of $\Phi_E$ in cm$^{-2}$ eV$^{-1}$.*

**Solution:**
A spectral fluence is differential in energy, and as the output corresponds to tracks integrated over the volume of the region, its units correspond to the product $V\Phi_E$, that is, cm$^3$/(cm$^2$ eV) = cm/eV. (Note that the reason for this peculiarity is that Penelope uses a geometry package based on quadric surfaces, and in general cannot determine the volume of a body during the simulation process.) Thus, the user must divide the Monte Carlo score output by the body volume to obtain the correct units for the fluence differential in energy:

$$\Phi_E = \frac{(V\ \Phi_E)_{MC}}{V_{user}} = \frac{cm/eV}{cm^3} = \frac{1}{cm^2\ eV}$$

**7** A point source isotropically emitting $10^8$ fast neutrons per second falls out of its shielding onto a railroad platform 3 m horizontally from the track. A train goes by at 60 km h$^{-1}$. Ignoring scattering and attenuation, what is the fluence of neutrons that would strike a passenger at the same height above the track as the source?

*Answer: 5 × 10$^5$ neutrons m$^{-2}$*

**Solution:**
See scheme in Figure 4.2. Let $t = 0$ be the time when the passenger passes over the point on the track closest to the source, that is, $x = 0$.
From the motion dynamics,

$$\frac{dx}{dt} = 60\ \frac{km}{h} \times 10^3\ \frac{m}{km} \times \frac{h}{3600\ s} = 16.67\ \frac{m}{s},$$

$$\therefore\ x\ (\text{in metres}) = 16.67\ t\ (\text{in seconds})$$

and from the figure

$$\ell^2 = x^2 + 3^2 = (16.67\ t)^2 + 9\ m^2$$

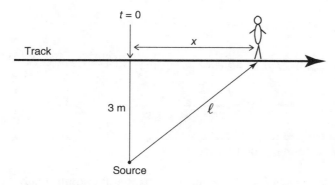

**Figure 4.2** Scheme for Exercise 7.

At a distance of $\ell$ meters from the source, the neutron fluence rate is

$$\dot{\Phi} = \frac{10^8}{4\,\pi\,\ell^2} = \frac{7.958\times10^6}{\ell^2}\frac{\text{neutrons}}{\text{m}^2\ \text{s}}$$

As a function of time, the neutron fluence rate striking the passenger is

$$\dot{\Phi}(t) = \frac{7.958\times10^6}{(16.67\ t)^2+9}\frac{\text{neutrons}}{\text{m}^2\ \text{s}}$$

and the fluence from $t=0$ to $\infty$ is

$$\Phi(0,\infty) = \int_0^\infty \frac{7.958\times10^6}{(16.67\ t)^2+9}\ dt$$

This integral can be solved numerically or using tables of integrals. For the latter, we find

$$\int_0^\infty \frac{a}{a^2+p^2}\ dp = \frac{\pi}{2}\quad\text{for } a>0$$

Thus we can write

$$\Phi(0,\infty) = \frac{7.958\times10^6}{(16.67)^2}\int_0^\infty \frac{1}{t^2+\frac{9}{(16.67)^2}}\ dt$$

$$= \frac{7.958\times10^6}{(16.67)^2}\frac{16.67}{3}\int_0^\infty \frac{\frac{3}{16.67}}{t^2+\frac{9}{(16.67)^2}}\ dt\ = 2.5\times10^5\frac{\text{neutrons}}{\text{m}^2},$$

which must be doubled to account also for the contribution from $t=-\infty$ to 0, giving

$$\Phi(-\infty,\infty) = 5\times10^5\ \text{neutrons m}^{-2}$$

**8** An x-ray field at a point $P$ contains $7.5\times10^8$ photons/(m$^2$ s keV), with energies uniformly distributed from 10 keV to 100 keV. (a) What is the photon fluence rate at $P$? (b) What would be the photon fluence in 1 h? (c) What is the corresponding energy fluence in J/m$^2$ and erg/cm$^2$?
*Answer:* *(a)* $6.75\times10^{10}$ *photons* $m^{-2}\,s^{-1}$; *(b)* $2.43\times10^{14}$ *photons* $m^{-2}$; *(c) 2.14 J* $m^{-2}$ *or* $2.14\times10^3$ *erg* $cm^{-2}$

**Solution:**

(a) The photon fluence rate is obtained as

$$\dot{\Phi} = \int_0^{E_{max}} \dot{\Phi}_E\ dE = \int_0^{10\ \text{keV}} 0\ dE$$

$$+ \int_{10\ \text{keV}}^{100\ \text{keV}} 7.5\times10^8\ \frac{\text{photons}}{\text{m}^2\ \text{s keV}}\ dE(\text{keV})$$

$$= 0 + 7.5\times10^8\ [100-10] = 6.75\times10^{10}\ \frac{\text{photons}}{\text{m}^2\ \text{s}}$$

(b) In 1 h the fluence is

$$\Phi(t, 1\,\text{h}) = \dot{\Phi}\,\Delta t = 6.75 \times 10^{10}\,\frac{\text{photons}}{\text{m}^2\,\text{s}}$$

$$\times\, 3600\,\frac{\text{s}}{\text{h}} = 2.43 \times 10^{14}\,\frac{\text{photons}}{\text{m}^2}$$

(c) Recalling that $\dot{\Psi} = E\,\dot{\Phi}$, the energy fluence rate is given by

$$\dot{\Psi} = \int_0^{E_{\max}} \dot{\Phi}_E\, E\, \mathrm{d}E = \int_0^{10\,\text{keV}} 0\, E\, \mathrm{d}E$$

$$+ \int_{10\,\text{keV}}^{100\,\text{keV}} 7.5 \times 10^8\,\frac{\text{photons}}{\text{m}^2\,\text{s keV}}\, E(\text{keV})\, \mathrm{d}E(\text{keV})$$

$$= 0 + \frac{7.5 \times 10^8}{2}[E^2]_{10}^{100} = 3.75 \times 10^8\,[10^4 - 10^2]$$

$$= 3.71 \times 10^{12}\,\frac{\text{keV}}{\text{m}^2\,\text{s}} \times 1.6022 \times 10^{-16}\,\frac{\text{J}}{\text{keV}}$$

$$= 5.95 \times 10^{-4}\,\frac{\text{J}}{\text{m}^2\,\text{s}}$$

and the energy fluence in 1 h is

$$\Psi = 5.95 \times 10^{-4}\frac{\text{J}}{\text{m}^2\,\text{s}} \times 3600\,\frac{\text{s}}{\text{h}} = 2.14\,\frac{\text{J}}{\text{m}^2} \times 10^7\,\frac{\text{erg}}{\text{J}} \times \frac{\text{m}^2}{10^4\,\text{cm}^2}$$

$$= 2.14 \times 10^3 \frac{\text{erg}}{\text{cm}^2}$$

**9**  Show that, for a spherical volume, Kellerer's definition of fluence (often cited as Chilton's) gives the same value as the conventional definition in a uniform monoenergetic field. *Hint*: The mean chord length in any convex volume is $\ell = 4v/a$, where $v$ is the volume and $a$ is the surface area.
*Answer: For a finite sphere of radius r, the fluence is given by*

$$\Phi = \frac{N}{\pi\,r^2}$$

*Kellerer's formulation is*

$$\Phi = \frac{N\ell}{v} = \frac{N\frac{4v}{a}}{v} = \frac{4N}{a} = \frac{N}{\pi\,r^2},$$

*which is identical to the expression above.*

**10**  What is $(K_{\text{el}})_{\text{air}}$ in gray at a point in air where $X = 47$ roentgens?
*Answer: 0.412 Gy*

**Solution:**
$(K_{\text{el}})_{\text{air}}[\text{Gy}] = X[\text{C/kg}]\,(W/e)_{\text{air}}$, with $(W/e)_{\text{air}} = 33.97\,\text{J/C}$
and $1\,\text{R} = 2.580 \times 10^{-4}\,\text{C/kg}$

Therefore

$$X = 47 \text{ R} \times \frac{2.58 \times 10^{-4} \text{ C/kg}}{1 \text{ R}} = 0.01213 \text{ C/kg}$$

$$(K_{\text{el}})_{\text{air}} = 0.01213 \times 33.97 = 0.412 \text{ Gy}$$

**11** An electron enters a volume $v$ with a kinetic energy of 4 MeV and carries 0.5 MeV of that energy out of $v$ when it leaves. While in the volume it produces a bremsstrahlung photon of 1.5 MeV, which escapes from $v$. What is the contribution of this event to (a) the energy transferred, (b) the net energy transferred, and (c) the energy imparted?
*Answer: (a) 0 MeV; (b) 0 MeV; (c) 2 MeV*

**Solution:**
See Figure 4.3.
(a) $\varepsilon_{\text{tr}} = (R_{\text{in}})_u - (R_{\text{out}})_u^{\text{non-r}} + \sum Q = 0 - 0 + 0 = 0$
where $(R_{\text{in}})_u = 0$ because no uncharged radiation enters $v$ and $(R_{\text{out}})_u^{\text{non-r}} = 0$ because the photon $k$ originated from a radiative loss by the electron while in $v$.
(b) $\varepsilon_{\text{tr}}^n = \varepsilon_{\text{tr}} - R_u^r = 0$
where $R_u^r = 0$ because the electron did not originate in $v$.
(c) $\varepsilon = (R_{\text{in}})_u - (R_{\text{out}})_u + (R_{\text{in}})_c - (R_{\text{out}})_c + \sum Q = 0 - 1.5 + 4 - 0.5 + 0 = 2$ MeV, which is the energy spent by the electron in electronic interactions in $v$.

**12** A 10 MeV $\gamma$ ray enters a volume $v$ and undergoes pair production, thereby disappearing and giving rise to an electron and positron of equal energies. The electron spends half its kinetic energy in collision interactions before escaping from $v$. The positron spends half of its kinetic energy in electronic collisions in $v$ before being annihilated in flight. The resulting photons escape from $v$. Determine (a), (b), and (c) as in the previous exercise.
*Answer: (a) 8.98 MeV; (b) 6.73 MeV; (c) 4.49 MeV*

**Solution:**
See Figure 4.4. The total energy carried out by annihilation photons, $k_1 + k_2 = 1.022$ MeV from rest mass $+ 2.245$ MeV final kinetic energy of positron $= 3.267$ MeV.

**Figure 4.3** An electron enters volume $v$ with a kinetic energy of 4 MeV and carries 0.5 MeV of that energy out of $v$ when it leaves; within the volume the electron produces a bremsstrahlung photon of 1.5 MeV that escapes from $v$.

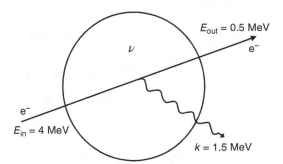

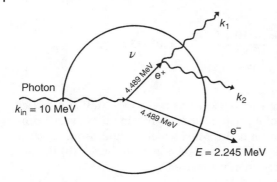

**Figure 4.4** A 10 MeV γ ray enters a volume $v$ and undergoes pair production, thereby disappearing and giving rise to an electron and positron of equal energies. The electron spends half its kinetic energy in electronic interactions before escaping from $v$. The positron spends half of its kinetic energy in collisions in $v$ before being annihilated in flight. The resulting photons escape from $v$.

(a)

$$\varepsilon_{tr} = (R_{in})_u - (R_{out})_u^{non-r} + \sum Q$$

$$= 10 - 1.022_{rest\ m} + (1.022_{annih} - 1.022_{pair\ prod}) = 8.978 \text{ MeV}$$

where $1.022_{rest\ m}$ is the part of the energy carried out by $k_1 + k_2$ that was obtained from rest mass in the positron–electron annihilation.

(b)

$$\varepsilon_{tr}^n = \varepsilon_{tr} - R_u^r = 8.978 - 2.245 = 6.733 \text{ MeV}$$

2.245 MeV is the part of the energy carried out by $k_1 + k_2$ that was obtained from the remaining kinetic energy of the positron at its annihilation; thus it is radiative loss by a charged particle that started in $v$.

(c) The energy imparted

$$\varepsilon = (R_{in})_u - (R_{out})_u + (R_{in})_c - (R_{out})_c + \sum Q$$

$$= 10 - 3.267 + 0 - 2.245 + (1.022 - 1.022) = 4.488 \text{ MeV}$$

**13** The fluence rate of 6 MeV γ rays is $3.5 \times 10^6$ photons cm$^{-2}$ s$^{-1}$ at a point in Pb. What are the values of $K$ and $K_{el}$ there after 1 week? (Express in units of erg/g and gray.)
*Answer:* $K = 7.65 \times 10^5$ erg g$^{-1}$ = 76.5 Gy; $K_{el} = 5.58 \times 10^5$ erg g$^{-1}$ = 55.8 Gy

**Solution:**

(a) For the total kerma, $K = \Psi\,(\mu_{tr}/\rho)_{Pb}$
where

$$\Psi = 3.5 \times 10^6\ \frac{phot}{cm^2\ s} \times 6.048 \times 10^5\ s \times \frac{6\ MeV}{phot} \times \frac{1.6022 \times 10^{-6}\ erg}{MeV}$$

$$= 2.0349 \times 10^7\ \frac{erg}{cm^2}$$

and from the Data Tables

$$(\mu_{tr}/\rho)_{Pb,6\ MeV} = 0.0376\ cm^2/g$$

Therefore

$$K = 2.0349 \times 10^7 \, \frac{erg}{cm^2} \times 0.0376 \, \frac{cm^2}{g} = 7.6513 \times 10^5 \, \frac{erg}{g} = 76.51 \, Gy$$

(b) For the electronic kerma, $K_{el} = \Psi \, (\mu_{en}/\rho)_{Pb}$

$$(\mu_{en}/\rho)_{Pb,6 \, MeV} = 0.0274 \, cm^2/g$$

Therefore

$$K_{el} = 2.0349 \times 10^7 \, \frac{erg}{cm^2} \times 0.0274 \, \frac{cm^2}{g} = 5.5757 \times 10^5 \, \frac{erg}{g} = 55.76 \, Gy$$

**14** A field of 14.5 MeV neutrons deposits a kerma of 1.37 Gy at a point of interest in water. What is the fluence?
*Answer: 2.015 × 10¹⁰ neutrons/cm²*

**Solution:**
For the neutron kerma, Eq. (4.68):

$$K_w = \Phi_w \, k_\Phi (E_n)_w = 1.37 \, Gy$$

where, from the Data Tables, the neutron *kerma coefficient* for 14.5 MeV in water is

$$k_\Phi = 6.80 \times 10^{-15} \, Gy \, m^2$$

Therefore

$$\Phi_w = \frac{K_w}{k_\Phi (E_n)_w} = \frac{1.37 \, Gy}{6.80 \times 10^{-15} \, Gy \, m^2 \times 10^4 \frac{cm^2}{m^2}} = 2.015 \times 10^{10} \, \frac{1}{cm^2}$$

**15** An x-ray field at a point $P$ in aluminium contains $7.5 \times 10^8$ photons m⁻²s⁻¹ keV⁻¹, uniformly distributed from 10 keV to 100 keV. Calculate the electronic kerma at $P$ for a 1 h irradiation, in Gy. (Note: Use log–log interpolation from the electronic Data Tables.)
*Answer: ~ 0.093 Gy*

**Solution:**
The equation for $\dot{K}_{el}$ for a spectrum (see Eq. (4.65) in the textbook) is

$$(\dot{K}_{el})_{Al} = \int_{10 \, keV}^{100 \, keV} [\dot{\Psi}_k]_{Al} \, [\mu_{en}(k)/\rho]_{Al} \, dk \approx \sum_i [\dot{\Psi}_k]_i \, [\mu_{en}(k)/\rho]_i \, [\Delta k]_i$$

with units

$$Gy \left[ \frac{J}{kg} \right] = \frac{J}{m^2 \, keV} \, \frac{m^2}{kg} \, keV$$

The energy fluence rate is

$$\dot{\Psi}_k = k \, \dot{\Phi}_k = k \, \frac{keV}{phot} \times 7.5 \times 10^8 \, \frac{phot}{m^2 \, s \, keV} \times 3600 \, \frac{s}{h}$$

$$\times\ 1.6022 \times 10^{-16}\ \frac{J}{keV}$$

$$=\ 4.325 \times 10^{-4}\ k\ \frac{J}{m^2\ h\ keV}$$

A numerical integration, based on log–log interpolation for $[\mu_{en}/\rho]_{Al}$ from the electronic Data Tables, would be as follows:

| k-Interval | k | $[\mu_{en}(k)/\rho]_{Al}$ | $\Delta k$ | $\dot{\Psi}(k)$ | $\dot{\Psi}(k)\ [\mu_{en}/\rho]\ \Delta k$ |
|---|---|---|---|---|---|
| | keV | m²/kg | keV | J/(m² keV h) | J/(kg h) |
| 10–15 | 12.5 | 1.287E+00 | 5 | $5.406 \times 10^{-3}$ | $3.478 \times 10^{-2}$ |
| 15–20 | 17.5 | 4.608E−01 | 5 | $7.569 \times 10^{-3}$ | $1.744 \times 10^{-2}$ |
| 20–30 | 25 | 1.528E−01 | 10 | $1.081 \times 10^{-2}$ | $1.652 \times 10^{-2}$ |
| 30–40 | 35 | 5.359E−02 | 10 | $1.514 \times 10^{-2}$ | $8.112 \times 10^{-3}$ |
| 40–50 | 45 | 2.497E−02 | 10 | $1.946 \times 10^{-2}$ | $4.860 \times 10^{-3}$ |
| 50–60 | 55 | 1.382E−02 | 10 | $2.379 \times 10^{-2}$ | $3.287 \times 10^{-3}$ |
| 60–80 | 70 | 7.341E−03 | 20 | $3.028 \times 10^{-2}$ | $4.445 \times 10^{-3}$ |
| 80–100 | 90 | 4.394E−03 | 20 | $3.893 \times 10^{-2}$ | $3.421 \times 10^{-3}$ |
| $\dot{K}_{el}$ | | | | | $\Sigma = 0.093$ Gy h$^{-1}$ |

A 'proper' numerical integration (here done with Mathematica) yields $\dot{K}_{el} = 0.096$ Gy h$^{-1}$.

**16** Consider two flasks containing 5 cm$^3$ and 25 cm$^3$ of water, respectively. They are identically and homogeneously irradiated with γ rays, making the average kerma equal to 1 Gy in the smaller flask. (a) Neglecting differences in γ-ray attenuation, what is the average kerma in the larger flask? (b) What is the energy transferred in each volume of water?
*Answer: (a) 1 Gy; (b) 0.005 J, 0.025 J*

**Solution:**

(a) $K = \Psi\ (\mu_{tr}/\rho)$ at each point, and since $\Psi$ is homogeneous and identical in both volumes, $K$ is equal to its average value at all points in both volumes, that is, $\bar{K} = 1$ Gy in the larger volume.

(b) From Eq. (4.44)

$$K = \frac{dE_{tr}}{dm}$$

Assuming unit density, the masses of water in the two flasks are 5 g and 25 g, or 0.005 kg and 0.025 kg. For these finite masses, uniformly irradiated, the energy transferred for the small volume is

$$E_{tr} = K\, m = 1 \text{ Gy} \times 0.005 \text{ kg} = 0.005 \text{ J}$$

and for the large volume is

$$1 \text{ Gy} \times 0.025 \text{ kg} = 0.025 \text{ J}$$

# 5

# Elementary Aspects of the Attenuation of Uncharged Particles through Matter

**1** Let $\mu_1 = 0.02$ cm$^{-1}$ and $\mu_2 = 0.04$ cm$^{-1}$ be the partial linear attenuation coefficients in the slab shown in Figure 5.1. Let $L = 5$ cm and $N_0 = 10^6$ particles. How many particles $N_L$ are transmitted, and how many are absorbed by each interaction process in the slab?

*Answer: $N_L = 7.41 \times 10^5$; $N_1 = 8.64 \times 10^4$; $N_2 = 1.73 \times 10^5$*

**Solution:**

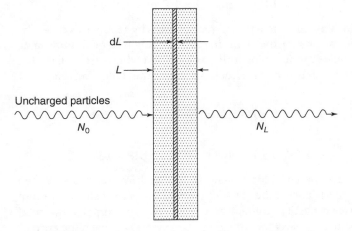

**Figure 5.1** Exponential attenuation of uncharged particles.

The number of transmitted particles $N_L$ is given by

$$N_L = N_0 \, e^{-(\mu_1 + \mu_2)L} = 10^6 \, e^{-(0.02 + 0.04)5} = 7.408 \times 10^5$$

The total number of particles absorbed is

$$N_0 - N_L = (10^6 - 7.408 \times 10^5) = 2.592 \times 10^5$$

The number absorbed by process 1 is

$$\Delta N_1 = (N_0 - N_L)\frac{\mu_1}{\mu} = 2.592 \times 10^5 \times \frac{0.02}{0.06} = 8.64 \times 10^4$$

*Fundamentals of Ionizing Radiation Dosimetry: Solutions to Exercises,* First Edition.
Pedro Andreo, David T. Burns, Alan E. Nahum, and Jan Seuntjens.

and by process 2,

$$\Delta N_2 = (N_0 - N_L)\frac{\mu_2}{\mu} = 2.592 \times 10^5 \times \frac{0.04}{0.06} = 1.728 \times 10^5$$

Note that in this exercise we cannot derive the number of process 1 events on the basis of $\mu_1$ alone, since the number of particles available for interaction at any depth in the slab depends on the total attenuation coefficient $\mu$. Confusion on this point sometimes results from the fact that in an *infinitesimal* layer the number of interactions by each process can be obtained from Eq. (5.1) in the textbook:

$$dN_1 = -\mu_1 N\, dL$$
$$dN_2 = -\mu_2 N\, dL$$

However, for non-infinitesimal layers this formula does not apply. If one attempts to use this equation to solve the exercise, one obtains

$$\Delta N_1 \neq -0.02\ \text{cm}^{-1} \times 10^6 \times 5\ \text{cm} = 1 \times 10^5\ \text{interactions}$$

by process 1 and

$$\Delta N_2 \neq -0.04\ \text{cm}^{-1} \times 10^6 \times 5\ \text{cm} = 2 \times 10^5\ \text{interactions}$$

by process 2, which overestimate the correct answers by 16% in this case. The thicker the layer, the greater the error due to assuming it to be infinitesimal. Another common error in doing this exercise is trying to calculate the number of individual process interactions from the following incorrect equations:

$$\Delta N_1 \neq N_0 - N_0\, e^{-\mu_1 L} = 10^6 - 10^6\, e^{-0.02\times 5} = 9.52 \times 10^4$$

and

$$\Delta N_2 \neq N_0 - N_0\, e^{-\mu_2 L} = 10^6 - 10^6\, e^{-0.04\times 5} = 1.813 \times 10^5,$$

which overestimate the correct answers by 10% and 5%, respectively. The latter answer is nearly correct because $\mu_2$ approximates the value of $\mu$ more closely. The closer a partial attenuation coefficient approximates the total coefficient $\mu$, the more nearly the foregoing statements approach the correct Eq. (5.9) in the textbook, that is,

$$\Delta N_x = (N_0 - N_L)\frac{\mu_x}{\mu} = N_0(1 - e^{-\mu L})\frac{\mu_x}{\mu}$$

**2** A plane-parallel monoenergetic beam of $10^{12}$ uncharged particles per second is incident perpendicularly on a layer of material 0.02 m thick, having a density $\rho = 11.3 \times 10^3\ \text{kg m}^{-3}$. For values of the mass attenuation coefficient $\mu/\rho = 1 \times 10^{-3}, 3 \times 10^{-4}$, and $1 \times 10^{-4}\ \text{m}^2\ \text{kg}^{-1}$, calculate the number of primary particles transmitted in 1 min. Compare in each case, giving percentage errors, with the approximation $N_L/N_0 = e^{-\mu L} \approx 1 - \mu L$. *Answer: 4.786 × 10¹³, 5.607 × 10¹³, 5.866 × 10¹³; 4.644 × 10¹³, 5.593 × 10¹³, 5.864 × 10¹³; 3%, 0.25%, and 0.03% low.*

**Solution:**
From Eq. (5.3)

$$\frac{N_L}{N_0} = e^{-\mu L} = e^{-\frac{\mu}{\rho}\,\rho L}$$

and for this exercise the values are

$$\rho L = 11.3 \times 10^3 \frac{\text{kg}}{\text{m}^3} \times 0.02 \text{ m} = 2.26 \times 10^2 \frac{\text{kg}}{\text{m}^2}$$

$$N_0 = 60 \text{ s} \times 10^{12} \frac{\text{particles}}{\text{s}} = 6 \times 10^{13} \text{ particles}$$

| $\mu/\rho$ | $e^{-\frac{\mu}{\rho}\rho L}$ | $N_0\, e^{-\frac{\mu}{\rho}\rho L}$ | $1 - \frac{\mu}{\rho}\rho L$ | $N_0\left(1 - \frac{\mu}{\rho}\,\rho L\right)$ | % Error |
|---|---|---|---|---|---|
| $1 \times 10^{-3}$ | 0.7977 | $4.786 \times 10^{13}$ | 0.7740 | $4.644 \times 10^{13}$ | 2.97 |
| $3 \times 10^{-4}$ | 0.9344 | $5.607 \times 10^{13}$ | 0.9322 | $5.593 \times 10^{13}$ | 0.24 |
| $1 \times 10^{-4}$ | 0.9777 | $5.866 \times 10^{13}$ | 0.9774 | $5.864 \times 10^{13}$ | 0.03 |

**3**  What is the mean free path in each case in Exercise 2?
*Answer: 0.0885 m, 0.295 m, 0.885 m*

**Solution:**

$$\mu_1 = 11.3 \times 10^3 \frac{\text{kg}}{\text{m}^3} \times 1 \times 10^{-3}\frac{\text{m}^2}{\text{kg}} = 11.3 \text{ m}^{-1}; \frac{1}{\mu_1} = 0.0885 \text{ m}$$

$$\mu_2 = 11.3 \times 10^3 \frac{\text{kg}}{\text{m}^3} \times 3 \times 10^{-4}\frac{\text{m}^2}{\text{kg}} = 3.39 \text{ m}^{-1}; \frac{1}{\mu_2} = 0.295 \text{ m}$$

$$\mu_3 = 11.3 \times 10^3 \frac{\text{kg}}{\text{m}^3} \times 1 \times 10^{-4}\frac{\text{m}^2}{\text{kg}} = 1.13 \text{ m}^{-1}; \frac{1}{\mu_3} = 0.885 \text{ m}$$

**4**  Suppose that the beam in Exercise 2 is attenuated simultaneously by three different processes having the given attenuation coefficients. (a) How many particles are transmitted in 1 min? (b) How many interactions take place by each process?
*Answer: (a) $4.373 \times 10^{13}$; (b) $1.162 \times 10^{13}, 3.49 \times 10^{12}, 1.16 \times 10^{12}$*

**Solution:**
(a) From Eq. (5.7)

$$\frac{N_L}{N_0} = e^{-\mu L} = e^{-(\mu_1 + \mu_2 + \mu_3)L}$$

$$\frac{\mu}{\rho} = 1 \times 10^{-3} + 3 \times 10^{-4} + 1 \times 10^{-4} = 1.4 \times 10^{-3}\frac{\text{m}^2}{\text{kg}}$$

$$N_L = 6 \times 10^{13} \times e^{-1.4 \times 10^{-3} \times 2.26 \times 10^2} = 4.373 \times 10^{13} \text{ particles}$$

(b) Total number of interactions (Eq. (5.8)):

$$\Delta N = N_0 - N_L = N_0 - N_0\, e^{-\mu L} = N_0(1 - e^{-\frac{\mu}{\rho}\rho L})$$
$$= 6 \times 10^{13}\, (1 - e^{-1.4\times 10^{-3}\times 2.26\times 10^2}) = 1.627 \times 10^{13}$$

The number of interactions by a single process $x$ alone is (Eq. (5.9))

$$\Delta N_x = (N_0 - N_L)\frac{\mu_x}{\mu} = N_0\,(1 - e^{-\frac{\mu}{\rho}\rho L})\frac{\mu_x}{\mu}$$

$$\Delta N_1 = 1.627 \times 10^{13} \times \frac{1\times 10^{-3}}{1.4\times 10^{-3}} = 1.162 \times 10^{13}$$

$$\Delta N_2 = 1.627 \times 10^{13} \times \frac{3\times 10^{-4}}{1.4\times 10^{-3}} = 3.49 \times 10^{12}$$

$$\Delta N_3 = 1.627 \times 10^{13} \times \frac{1\times 10^{-4}}{1.4\times 10^{-3}} = 1.162 \times 10^{12}$$

**5** Suppose that a beam of uncharged radiation consists of one-third of particles of energy 2 MeV, for which $\mu/\rho = 1 \times 10^{-3}$ m$^2$/kg; one-third of 5 MeV particles, with $\mu/\rho = 3 \times 10^{-4}$ m$^2$/kg; and one-third of 7 MeV particles, with $\mu/\rho = 1 \times 10^{-4}$ m$^2$/kg. (a) What average value $\overline{(\mu/\rho)}_\Phi$ will be observed by a particle counter when a thin layer of the attenuator is interposed in the beam, with narrow-beam geometry? (b) Calculate the average $\overline{(\mu/\rho)}_\Psi$ that will be seen by an energy fluence meter.
*Answer: (a) 4.67 × 10$^{-4}$ m$^2$ kg$^{-1}$; (b) 3.00 × 10$^{-4}$ m$^2$ kg$^{-1}$*

**Solution:**
(a) Equation (5.12) rewritten for weighting by fluence $\Phi$ and divided by $\rho$ becomes

$$\overline{(\mu/\rho)}_{\Phi,L} = \frac{\int_{k=0}^{k_{max}} \Phi_L(k)\,[\mu(k)/\rho]_Z\,dk}{\int_{k=0}^{k_{max}} \Phi_L(k)\,dk}$$

$$= \frac{\sum_{i=1}^{3} (\Phi_i)_L[\mu(k)/\rho]_i}{\sum_{i=1}^{3} (\Phi_i)_L} = \frac{\sum\left[\frac{N_0}{3}\frac{\mu_1}{\rho} + \frac{N_0}{3}\frac{\mu_2}{\rho} + \frac{N_0}{3}\frac{\mu_3}{\rho}\right]}{\sum\left[\frac{N_0}{3} + \frac{N_0}{3} + \frac{N_0}{3}\right] = N_0}$$

$$= \frac{\frac{1}{3}\times 1 \times 10^{-3} + \frac{1}{3}\times 3 \times 10^{-4} + \frac{1}{3}\times 1 \times 10^{-4}}{1}$$

$$= 4.67 \times 10^{-4}\,\frac{m^2}{kg}$$

(b) For the energy fluence average

$$\overline{(\mu/\rho)}_{\Psi,L} = \frac{\int_{k=0}^{k_{max}} \Psi_L(k)\,[\mu(k)/\rho]_Z\,dk}{\int_{k=0}^{k_{max}} \Psi_L(k)\,dk}$$

$$= \frac{\sum_{i=1}^{3} (\Psi_i)_L \left(\frac{\mu}{\rho}\right)_i}{\sum_{i=1}^{3} (\Psi_i)_L} = \frac{\sum \left[\frac{E_1 N_0}{3} \frac{\mu_1}{\rho} + \frac{E_2 N_0}{3} \frac{\mu_2}{\rho} + \frac{E_3 N_0}{3} \frac{\mu_3}{\rho}\right]}{\sum \left[\frac{E_1 N_0}{3} + \frac{E_2 N_0}{3} + \frac{E_3 N_0}{3}\right]}$$

$$= \frac{\frac{2N_0}{3} \times 1 \times 10^{-3} + \frac{5N_0}{3} \times 3 \times 10^{-4} + \frac{7N_0}{3} \times 1 \times 10^{-4}}{\frac{2N_0}{3} + \frac{5N_0}{3} + \frac{7N_0}{3}}$$

$$= \frac{1.4 \times 10^{-3}}{4.667} = 3.00 \times 10^{-4} \frac{m^2}{kg}$$

**6** Let the beam in Exercise 5 first pass through a layer of the attenuator $250 \, kg \, m^{-2}$ thick in narrow-beam geometry. Then repeat (a) and (b).
*Answer: (a) $4.31 \times 10^{-4} \, m^2/kg$; (b) $2.79 \times 10^{-4} \, m^2/kg$*

**Solution:**
After the layer there are fewer particles and their proportions are changed:

For 2 MeV particles: $(N_L)_2 = \frac{N_0}{3} e^{-250 \times 1 \times 10^{-3}}$ $= 0.2596 \, N_0$

For 5 MeV particles: $(N_L)_5 = \frac{N_0}{3} e^{-250 \times 3 \times 10^{-4}}$ $= 0.3092 \, N_0$

For 7 MeV particles: $(N_L)_7 = \frac{N_0}{3} e^{-250 \times 1 \times 10^{-4}}$ $= 0.3251 \, N_0$

Total $= 0.8939 \, N_0$

Therefore, using the expressions in Exercise 5
(a) For the fluence-weighted average value

$$\overline{(\mu/\rho)}_\Phi = \frac{0.2596 \times 1 \times 10^{-3} + 0.3092 \times 3 \times 10^{-4} + 0.3251 \times 1 \times 10^{-4}}{0.8939}$$

$$= 4.31 \times 10^{-4} \frac{m^2}{kg}$$

(b) For the energy fluence average value

$$\overline{(\mu/\rho)}_\Psi$$
$$= \frac{2 \times 0.2596 \times 1 \times 10^{-3} + 5 \times 0.3092 \times 3 \times 10^{-4} + 7 \times 0.3251 \times 1 \times 10^{-4}}{2 \times 0.2596 + 5 \times 0.3092 + 7 \times 0.3251}$$

$$= 2.79 \times 10^{-4} \frac{m^2}{kg}$$

**7** At a depth of 47 cm in a medium, the absorbed dose is found to be 3.95 Gy, while that resulting only from primary radiation is 3.40 Gy. At the front surface of the medium, the dose from primary radiation is 10.0 Gy. Calculate the dose buildup factor $B$, the linear attenuation coefficient $\mu$, and the mean effective attenuation coefficient $\mu_{eff}$. Assume CPE and plane, monoenergetic primaries.
*Answer: $B = 1.162$, $\mu = 2.295 \, m^{-1}$, $\mu_{eff} = 1.976 \, m^{-1}$*

**Solution:**

Equation (5.3) rewritten for dose from primaries only is

$$\frac{D_L}{D_0} = \frac{3.40}{10.0} = e^{-\mu \times 0.47 \, \text{m}} = 0.34$$

from where

$$\mu = \frac{-\ln 0.34}{0.47} = \frac{1.079}{0.47} = 2.295 \, \text{m}^{-1}$$

For broad beam (prim + sec), Eq. (5.14) rewritten for dose is

$$\frac{D_L}{D_0} = B \, e^{-\mu \times 0.47} = 0.34,$$

that is,

$$B = \frac{0.395}{0.34} = 1.162,$$

and

$$0.395 = e^{-\mu_{\text{eff}} \times 0.47}$$

that is,

$$\mu_{\text{eff}} = 1.976 \, \text{m}^{-1}$$

# 6

# Macroscopic Aspects of the Transport of Radiation through Matter

Using any of the Monte Carlo systems described in Chapter 8, the reader should reproduce some of the figures of this chapter.

As there seems to be a widespread misunderstanding about how the radiation source descriptions for pencil, plane-parallel, and divergent beams should be defined by the user for Monte Carlo calculations based on the *EGSnrc* or *PENELOPE* user codes available, as well as on how the resulting quantities should be normalized, some clarifications are given in what follows.

In all cases, the source description is closely related to the scoring geometry so that we cannot refer to one omitting the other:

(1) A *pencil beam* is a point radiation source (zero area) that impinges on the surface of an infinitely broad phantom, and quantities are scored in infinitely broad slabs. A calculated broad beam depth-dose distribution, for example, will correspond to that of the central axis of an infinitely broad beam, as a result of the *reciprocity principle* (see Figure 6.1).

(2) A *plane-parallel beam* is, as inferred from its name, a parallel radiation source of a given width or radius that impinges on the surface of a phantom of certain dimensions. Obviously it should not be confused with a *divergent beam*, which is a point source at a given distance from the phantom that creates a conical or a pyramidal beam of a certain radius or dimensions at the surface. Quantities can be scored in any geometrical configuration, for example, in a narrow central-axis cylinder or in annuli of different radii.

(3) A *broad plane-parallel beam* is a plane-parallel beam of infinite dimensions. Its scoring requires special attention because of the following:

   (a) If scoring is done in a narrow central-axis cylinder, then the depth distribution will be identical to that of a pencil beam due to the *reciprocity principle* (see Figure 6.1)

   (b) If scoring is done in regions broader than the narrow central-axis cylinder, then no reciprocity is possible. The result will simply correspond to the defined scoring geometry, for example, a 'shower' of neutrons or photons in a nuclear reactor where the dose in a certain region is desired, or a photon radiotherapy treatment with a very large beam.

*Fundamentals of Ionizing Radiation Dosimetry: Solutions to Exercises*, First Edition.
Pedro Andreo, David T. Burns, Alan E. Nahum, and Jan Seuntjens.
© 2017 Wiley-VCH Verlag GmbH & Co. KGaA. Published 2017 by Wiley-VCH Verlag GmbH & Co. KGaA.

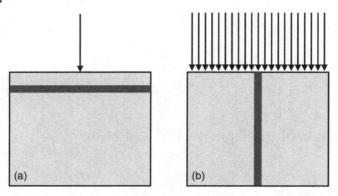

**Figure 6.1** Reciprocity principle in a Monte Carlo simulation: (a) using a pencil beam configuration and scoring a quantity irrespective of the phantom radius (up to infinity) is identical to (b) using an infinitely broad-parallel beam and scoring the quantity in a narrow central-axis cylinder.

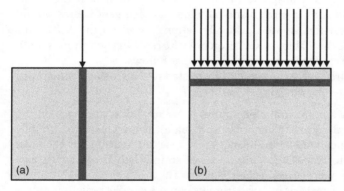

**Figure 6.2** These examples do NOT correspond to the reciprocity principle. The two configurations shown [as well as (b) with a smaller beam radius] will lead to completely different results, no matter how the normalization is made, as lateral scattering is different in the two cases.

*Hence, the reciprocity principle is only applicable to Cases 1 and 3a.* It should be emphasized that Case 1 is much more efficient than Case 3a as the scoring is done over the entire extension of the scoring slab.

Figure 6.2 illustrates configurations that *cannot* be considered to satisfy the reciprocity principle.

## Implementation

Implementing the cases above is in principle simpler in *PENELOPE* than in *EGSnrc* because sources are generally characterized by a cone or a pyramid of given aperture angle (setting it to zero yields a pencil beam) and the rest is done with a scoring geometry according to the descriptions above.

For *EGSnrc*, many user codes[1] are based on a set of pre-defined source geometries, of which source numbers 0, 1, and 2 are related to the discussion above. Because they implement some default values, some care is needed in certain cases:

(1) Source no. 0. Described as *Parallel beam incident from front (+Z-axis)*, where the beam radius is defined by the user (as well as the directional cosines).
In principle, this can be used to generate pencil beam distributions equivalent to those of a broad beam by setting the radius equal to zero, but care is needed because it *defaults to the max radius of the geometry*. This means that if the central-axis distribution of a broad beam is required, then at least *two scoring radial zones must be use*d: one for the central axis and the other infinitely broad. In the latter case, source no. 0 corresponds to Case 3a.
(2) Source no. 1. Described as *Point source on axis incident from front*, where the beam radius at the phantom surface and the source-to-surface distance are defined by the user.
There should not be any problem with this configuration, identical to the conical beam case mentioned above. The scoring geometry will be that described by the user for the central axis or annuli of different radii.
(3) Source no. 2. Described as *Broad parallel beam from front (+Z-axis)*, where no parameters are needed.
This is the case corresponding to the true pencil beam (Case 1), that is always much more efficient than source no. 0 for the reasons already given.

## Normalization of Results

It is also of interest to refer to the normalization of results in both sets of user codes because sometimes 'unconventional' conversions are made.
Recall first that $Gy = J/kg$; $MeV = 1.6022 \times 10^{-13}$ J; $kg = 10^3$ g.

(1) *EGSnrc* user codes give results in Gy *per incident fluence* ($Gy\ cm^2$) or fluence *per incident fluence* (dimensionless). Units of Gy are easy to convert to/from, for example, MeV/g, but some remarks are still required:
(a) For source no. 0, the incident fluence will be the number of histories divided by the beam area ($\Phi_0$ = # hist $cm^{-2}$), that is, the absorbed dose in $Gy\ cm^2$ is obtained from the energy deposition per mass of the scoring volume divided by the incident fluence:

$$\frac{\frac{MeV}{kg}}{\frac{\#\ hist}{cm^2}} = \frac{MeV}{kg}\frac{cm^2}{\#\ hist} = Gy\ cm^2$$

The fluence differential in energy ($MeV^{-1}\ cm^{-2}$) per incident fluence ($cm^{-2}$) has units of $MeV^{-1}$.
(b) For source no. 1, the same considerations apply.

---

1 Usually based on cylindrical geometry, but this is not relevant to the discussion.

This is not entirely correct as it does not take into account that the incident fluence is not parallel (a point-source beam is divergent). Some user codes (e.g., *egs_chamber*) partly compensate for this effect, replacing the beam area by an approximation to the solid angle subtended by the beam, that is, units will become per stereo radian.

(c) For source no. 2, the incident fluence is taken to be equal to the number of particles, as the beam area is infinite.

It is difficult to understand that results for this configuration involve units of Gy at all, as the mass of any region would also be infinite.

A physically correct normalization would be to the thickness (in g cm$^{-2}$) of the scoring slab, that is, express energy deposition per thickness in MeV cm$^2$ g$^{-1}$ (same units as the mass stopping power), and fluence per thickness in MeV$^{-1}$ g$^{-1}$.

All these differences, depending on the source number used, need to be taken into account for any comparison.

(2) *PENELOPE* user codes usually give results in terms of energy deposited or fluence within a slab *per incident particle*. The use of quadric surfaces to describe geometrical regions does not allow determining the volume of a region during the geometry-processing procedure (except when the scoring geometry is cylindrical), and any normalization per volume, per mass, or per thickness must be done a posteriori.

(a) 'Dose' calculations will be output in eV per history or eV/g per history for cylinders.

(b) Fluence calculations differential in energy are given in terms of 'fluence integrated over the region volume', in units of eV$^{-1}$ cm. This is a concept that eliminates the limitation of not knowing the volume during the simulation as

$$\Phi = \frac{\sum \text{tracks}}{\text{volume}} = \frac{\text{cm}}{\text{cm}^3} = \text{cm}^{-2}$$

$$\Phi \times \text{volume} = \text{cm}$$

$$\Phi_E \text{ has units of eV}^{-1} \text{ cm}^{-2}$$

$$\Phi_E \times \text{volume} = \text{eV}^{-1} \text{ cm}$$

This requires the user to divide the MC output by the volume of the region where the fluence differential in energy is calculated so that the units become eV$^{-1}$ cm$^{-2}$.

# 7

# Characterization of Radiation Quality

1   A photon spectrum is given with the unit of $m^{-2}$. Describe what one would interpret that such spectrum corresponds to, with respect to the units of the dosimetric quantity, justifying the answer. Clarify how misunderstandings in this respect (spectral quantity and units) can be avoided.
*Answer:*

If one were shown a plot of the spectrum with appropriate axis labels, there could be no confusion because a spectrum must always be given in terms of a quantity that is differential in energy, be it fluence differential in energy, $\Phi_E$, or energy fluence differential in energy, $\Psi_E$. That is, the plot of a spectrum indicates the following:

−The number of particles per unit area at each energy for the case of $\Phi_E$, that is, $[m^{-2}\ MeV^{-1}]$

−The energy transported by the particles per unit area at each energy for the case of $\Psi_E$, that is, $\Psi_E = E\ \Phi_E = [MeV \times m^{-2}\ MeV^{-1}] = [m^{-2}]$

This means that a spectrum in units of $m^{-2}$ can only be in terms of energy fluence.

The potential confusion could only arise because the total (integrated) fluence has also the units of $m^{-2}$, that is,

$$\Phi = \int_0^{E_{max}} \Phi_E\ dE = [m^{-2}\ MeV^{-1} \times MeV] = [m^{-2}],$$

but this could never correspond to a spectrum, that is, a series of values of a given quantity as a function of $E$. It would be the integrated spectrum, just one number (with a unit). The corresponding total energy fluence would be

$$\Psi = \int_0^{E_{max}} \Psi_E\ dE = \int_0^{E_{max}} E\ \Phi_E\ dE$$

$$= [MeV \times m^{-2}\ MeV^{-1} \times MeV] = [MeV\ m^{-2}],$$

which is again only one number (with a unit).

*Fundamentals of Ionizing Radiation Dosimetry: Solutions to Exercises,* First Edition.
Pedro Andreo, David T. Burns, Alan E. Nahum, and Jan Seuntjens.
© 2017 Wiley-VCH Verlag GmbH & Co. KGaA. Published 2017 by Wiley-VCH Verlag GmbH & Co. KGaA.

One way to avoid potential conflicts would be to use joules in the energy fluence to convey the sense of energy transported, that is,

$$\Psi_E = E\,\Phi_E = [J \times m^{-2}\ MeV^{-1}]$$

$$\Psi = \int_0^{E_{max}} E\,\Phi_E\ dE = [J \times m^{-2}\ MeV^{-1} \times MeV] = [J\ m^{-2}]$$

2   When a spectrum is not known, but only its attenuation properties can be measured, the parameter *energy equivalent* is sometimes used for x-ray beam quality specification. Give its definition and how it is determined. Apply the concept to determine the energy equivalent of an x-ray beam having an experimentally determined $HVL_1$ of 4 mm in aluminium.
*Answer: 38 keV*

**Solution:**
The *equivalent photon energy*, $k_{eq}$, is defined as the energy of a monoenergetic photon having the same $HVL_1$ as the beam being specified. In practice the value of $k_{eq}$ can be obtained by making use of the definition of $HVL_1$ in terms of air kerma:

$$\frac{K_{air}(HVL_1)}{K_{air}(0)} = \frac{1}{2} = e^{-(\mu/\rho)_{eq} HVL_1\,\rho}$$

that is,

$$(\mu/\rho)_{eq} = \frac{\ln\,2}{\rho\,HVL_1}\ cm^2\ g^{-1}$$

The value of $k_{eq}$ corresponding to $(\mu/\rho)_{eq}$ is then obtained by interpolation from $\mu(k)/\rho$ data, including Rayleigh (coherent) scattering.
For a $HVL_1 = 4$ mm in Al, for which $\rho = 2.6989$ g cm$^{-3}$

$$(\mu/\rho)_{eq} = \frac{0.69315}{2.6989 \times 0.4} = 0.6421\ cm^2\ g^{-1}$$

which by interpolation from the electronic Data Tables (Z13_aluminium_mutren.data) corresponds to an equivalent energy $k_{eq} = 37.81$ keV.

3   Using *SpekCalc* or a similar software to calculate kV x-ray spectra, calculate the fluence spectrum for 100 kV x rays for electron incidence on a 30° tungsten target, filtered by 3.5 mm of aluminium and with an inherent filtration of 3 mm beryllium; use 5 keV bins and a distance of 100 cm in air. (a) Determine the fluence-weighted, energy fluence-weighted, and air kerma-weighted mean energy of the spectrum, $\bar{k}$. (b) From the electronic Data Tables, estimate the corresponding mass energy-absorption coefficients of water and air and their ratio $[\mu_{en}(\bar{k})/\rho]_{w,air}$.
*Answer:*
*(a) $\bar{k}_\Phi = 50.25$ keV, $\bar{k}_\Psi = 56.17$ keV, and $\bar{k}_{K_{air}} = 43.47$ keV.*
*(b) $[\mu_{en}(\bar{k})/\rho]_{w,air} = 1.033,\ 1.042,$ and $1.023$, respectively.*

**Solution:**

The spectrum with 5 keV bins calculated using *SpekCalc* is shown in Figure 7.1 (solid line), where for comparison the spectrum calculated with a bin resolution of 1 keV (dashed line) is included. The latter shows more detail, but calculations with both bin widths yield similar results for the purposes of the exercise, while the 5 keV resolution spectrum is more suitable for simple numerical calculations. The numerical data for the fluence in each energy bin ($\Phi_k$) are given in the table below, where for convenience the various quantities to be used are also given. The symbol $k$ is used for the photon energy. The last two columns, for $[\mu_{en}(k)/\rho]_{air}$ and $[\mu_{en}(k)/\rho]_w$, are interpolated at each energy from the electronic Data Tables. Note that at the energies involved we can take $\mu_{en} = \mu_{tr}$, that is, $(1 - g) \approx 1$ because the electron radiation yield is practically negligible.

| | | | | | $K_{air,k} = k\,\Phi_k$ | | |
|---|---|---|---|---|---|---|---|
| Bin # | $k$ (keV) | $\Delta k$ (keV) | $\Phi_k$ (keV$^{-1}$ cm$^{-2}$) | $\Psi_k = k\,\Phi_k$ (J keV$^{-1}$ cm$^{-2}$) | $[\mu_{en}(k)/\rho]_{air}$ (J g$^{-1}$ keV$^{-1}$) | $[\mu_{en}(k)/\rho]_{air}$ (cm$^2$ g$^{-1}$) | $[\mu_{en}(k)/\rho]_w$ (cm$^2$ g$^{-1}$) |
| 1 | 10 | 5 | 2.314E−04 | 2.314E−03 | 1.070E−02 | 4.6251E+00 | 4.8191E+00 |
| 2 | 15 | 5 | 6.008E+03 | 9.012E+04 | 1.173E+05 | 1.3014E+00 | 1.3408E+00 |
| 3 | 20 | 5 | 4.784E+05 | 9.569E+06 | 5.028E+06 | 5.2542E−01 | 5.3719E−01 |
| 4 | 25 | 5 | 1.943E+06 | 4.859E+07 | 1.272E+07 | 2.6191E−01 | 2.6672E−01 |
| 5 | 30 | 5 | 3.184E+06 | 9.553E+07 | 1.440E+07 | 1.5074E−01 | 1.5217E−01 |
| 6 | 35 | 5 | 3.683E+06 | 1.289E+08 | 1.239E+07 | 9.6114E−02 | 9.7540E−02 |
| 7 | 40 | 5 | 3.656E+06 | 1.462E+08 | 9.810E+06 | 6.7084E−02 | 6.8235E−02 |
| 8 | 45 | 5 | 3.374E+06 | 1.518E+08 | 7.784E+06 | 5.1271E−02 | 5.2579E−02 |
| 9 | 50 | 5 | 3.001E+06 | 1.500E+08 | 6.048E+06 | 4.0313E−02 | 4.1644E−02 |
| 10 | 55 | 5 | 2.608E+06 | 1.435E+08 | 4.962E+06 | 3.4585E−02 | 3.5969E−02 |
| 11 | 60 | 5 | 3.947E+06 | 2.368E+08 | 7.122E+06 | 3.0069E−02 | 3.1467E−02 |
| 12 | 65 | 5 | 2.252E+06 | 1.464E+08 | 4.058E+06 | 2.7723E−02 | 2.9298E−02 |
| 13 | 70 | 5 | 1.603E+06 | 1.122E+08 | 2.886E+06 | 2.5714E−02 | 2.7424E−02 |
| 14 | 75 | 5 | 1.248E+06 | 9.359E+07 | 2.318E+06 | 2.4764E−02 | 2.6579E−02 |
| 15 | 80 | 5 | 1.005E+06 | 8.042E+07 | 1.923E+06 | 2.3908E−02 | 2.5812E−02 |
| 16 | 85 | 5 | 7.765E+05 | 6.600E+07 | 1.553E+06 | 2.3524E−02 | 2.5564E−02 |
| 17 | 90 | 5 | 5.569E+05 | 5.012E+07 | 1.161E+06 | 2.3168E−02 | 2.5332E−02 |
| 18 | 95 | 5 | 3.305E+05 | 3.140E+07 | 7.280E+05 | 2.3182E−02 | 2.5340E−02 |
| 19 | 100 | 5 | 0 | 0 | 0 | 2.3195E−02 | 2.5348E−02 |

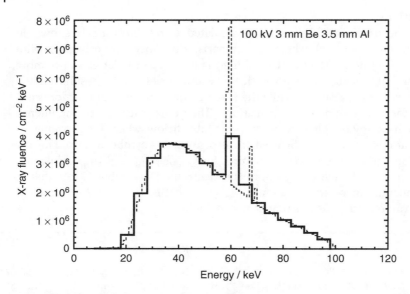

**Figure 7.1** Spectra of 100 kV x rays filtered by 3.5 mm of aluminium and inherent filtration of 3 mm beryllium using different resolutions, 5 keV (solid line) and 1 keV (dashed line). Calculations made with *SpekCalc*.

From the table, we get

$$\Phi = \sum \Phi_k \, \Delta k = 1.683 \times 10^8 \text{ cm}^{-2}$$

$$\Psi = \sum \Psi_k \, \Delta k = 8.456 \times 10^9 \text{ J cm}^{-2}$$

$$K_{\text{air}} = \sum K_{\text{air},k} \, \Delta k = 4.751 \times 10^8 \text{ J g}^{-1}$$

The different spectrum-weighted mean energies are given by

$$\bar{k}_\Phi = \frac{\sum k \, \Phi_k \, \Delta k}{\sum \Phi_k \, \Delta k} = \frac{8.456 \times 10^9}{1.683 \times 10^8} = 50.25 \text{ keV}$$

$$\bar{k}_\Psi = \frac{\sum k \, \Psi_k \, \Delta k}{\sum \Psi_k \, \Delta k} = \frac{4.750 \times 10^{11}}{8.456 \times 10^9} = 56.17 \text{ keV}$$

$$\bar{k}_{K_{\text{air}}} = \frac{\sum k \, K_{\text{air, k}} \, \Delta k}{\sum K_{\text{air, k}} \Delta k} = \frac{2.065 \times 10^{10}}{4.751 \times 10^8} = 43.47 \text{ keV}$$

and interpolating from the Data Tables for the $\mu_{\text{en}}(k)/\rho$ values, we get for each mean energy

| | $\bar{k}_\Phi$ <br> 50.25 keV | $\bar{k}_\Psi$ <br> 56.17 keV | $\bar{k}_{K_{\text{air}}}$ <br> 43.47 keV |
|---|---|---|---|
| $[\mu_{\text{en}}(\bar{k})/\rho]_{\text{w}} =$ | $4.132 \times 10^{-2}$ | $3.482 \times 10^{-2}$ | $5.675 \times 10^{-2} \text{ cm}^2 \text{ g}^{-1}$ |
| $[\mu_{\text{en}}(\bar{k})/\rho]_{\text{air}} =$ | $3.999 \times 10^{-2}$ | $3.343 \times 10^{-2}$ | $5.547 \times 10^{-2} \text{ cm}^2 \text{ g}^{-1}$ |
| $[\mu_{\text{en}}(\bar{k})/\rho]_{\text{w,air}} =$ | $1.033$ | $1.042$ | $1.023$ |

4  Using the spectrum calculated in the previous exercise, determine the fluence-weighted, energy fluence-weighted, and air kerma-weighted mean ratios of mass energy-absorption coefficients water to air. Compare with the value of the ratios of the mean coefficients in each case. Note that neither of these evaluations use the mean energy evaluted in the previous exercise.
   *Answer:*  $[(\bar{\mu}_{en}/\rho)_{w,air}]_\Phi = 1.036$; $[(\bar{\mu}_{en}/\rho)_{w,air}]_\Psi = 1.044$; $[(\bar{\mu}_{en}/\rho)_{w,air}]_{K_{air}} = 1.030$. *Differences with the ratios of the mean values are* $-1.2\%, -1.4\%$, *and* $-0.9\%$, *respectively.*

   **Solution:**
   Similar to the mean energies in the previous exercise, spectrum-weighted mean values of photon coefficients are given by

   $$(\bar{\mu}/\rho)_\Phi = \frac{\sum[\mu(k)/\rho]\Phi_k\Delta k}{\sum \Phi_k\Delta k}; \quad (\bar{\mu}/\rho)_\Psi = \frac{\sum[\mu(k)/\rho]\Psi_k\Delta k}{\sum \Psi_k\Delta k};$$

   $$(\bar{\mu}/\rho)_{K_{air}} = \frac{\sum[\mu(k)/\rho]K_{air,\,k}\Delta k}{\sum K_{air,\,k}\Delta k}$$

   and the weighted mean ratios of the coefficients are, for example, for the fluence-weighted case,

   $$[(\bar{\mu}/\rho)_{w,air}]_\Phi = \frac{\sum[\mu(k)/\rho]_{w,air}\Phi_k\Delta k}{\sum \Phi_k\Delta k}$$

   Note that the above is not the same as the ratio of the weighted mean values of the coefficients

   $$\frac{[(\bar{\mu}/\rho)_w]_\Phi}{[(\bar{\mu}/\rho)_{air}]_\Phi} = \frac{\frac{\sum[\mu(k)/\rho]_w\Phi_k\Delta k}{\sum \Phi_k\Delta k}}{\frac{\sum[\mu(k)/\rho]_{air}\Phi_k\Delta k}{\sum \Phi_k\Delta k}} = \frac{\sum[\mu(k)/\rho]_w\Phi_k\Delta k}{\sum[\mu(k)/\rho]_{air}\Phi_k\Delta k}$$

   Using the table of data from the previous exercise, for the weighted mean ratios of the coefficients, we have in each case

   $$[(\bar{\mu}_{en}/\rho)_{w,air}]_\Phi = \frac{\sum[\mu(k)/\rho]_{w,air}\Phi_k\Delta k}{\sum \Phi_k\Delta k} = \frac{1.744 \times 10^8}{1.683 \times 10^8} = 1.036$$

   $$[(\bar{\mu}_{en}/\rho)_{w,air}]_\Psi = \frac{\sum[\mu(k)/\rho]_{w,air}\Psi_k\Delta k}{\sum \Psi_k\Delta k} = \frac{8.828 \times 10^9}{8.456 \times 10^9} = 1.044$$

   $$[(\bar{\mu}_{en}/\rho)_{w,air}]_{K_{air}} = \frac{\sum[\mu(k)/\rho]_{w,air}K_{air,\,k}\Delta k}{\sum K_{air,\,k}\Delta k} = \frac{4.892 \times 10^8}{4.751 \times 10^8} = 1.030$$

   These values can be compared with the approximate results obtained in the previous exercise, calculated for the mean energies of the spectrum. They all agree quite well and only the air kerma case differs by 0.7%.
   The respective ratios of the weighted mean values are 1.024, 1.030, and 1.020, that is, differences with the weighted mean ratios of $-1.2\%, -1.4\%$, and $-0.9\%$, respectively.

5  Using the spectrum calculated in Exercise 3; (a) determine the $HVL_1$ in aluminium. (b) Determine the equivalent photon energy, $k_{eq}$, of the

spectrum, and from the elecronic Data Tables estimate the ratio of mass energy-absorption coefficients water to air at $k_{eq}$. (c) Compare the water/air ratios with the previous two exercises. *Hint: Recall the* $HVL_1$ *definition in terms of air kerma, that is,* $K_{air}(HVL_1)/K_{air}(0) = 1/2$. *For the fluence spectrum* $\Phi_k$, *build a table of* $K_{air}(0)$ *and* $K_{air}$(attenuated). *By iteration, vary HVL until the required ratio is found (it is suggested to use software for this step, e.g., Excel/Solver or MatLab).*
Answer: $HVL_1 = 3.8$ mm Al, $k_{eq} = 37$ keV, $[\mu_{en}(k_{eq})/\rho]_{w,air} = 1.016$

**Solution:**

(a) Recall that for a given fluence spectrum $\Phi_k$, the (integrated) air kerma is
$K_{air} = \sum K_{air,k} \, \Delta k$, with $K_{air,k} = k \, \Phi_k [\mu_{en}(k)/\rho]_{air}$ (see the table of data in Exercise 3). The air kerma attenuated by an aluminium thickness equal to the $HVL_1$ will be given by

$$K_{air}(HVL_1) = \sum K_{air,k} \, e^{-\mu(k) \times HVL_1} \, \Delta k$$

We now build a table of values for the non-attenuated air kerma $K_{air}(0)$ and for the attenuated air kerma $K_{air}(HVL_1)$ using a starting value for $HVL_1$, of, say, 0.5 cm of aluminium (1.3495 g cm$^{-2}$), as illustrated below. The second last column, $K_{air}(0.5)$, corresponds to this quantity.

| Bin # | k (keV) | Δk (keV) | $\Phi_k$ (keV$^{-1}$ cm$^{-2}$) | $[\mu_{en}(k)/\rho]_{air}$ (cm$^2$ g$^{-1}$) | $K_{air,k}(0)$ (J g$^{-1}$ keV$^{-1}$) | $[\mu(k)/\rho]_{Al}$ (cm$^2$ g$^{-1}$) | $K_{air,k}(0.5)$ (J g$^{-1}$ keV$^{-1}$) | $K_{air,k}(0.3803)$ (J g$^{-1}$ keV$^{-1}$) |
|---|---|---|---|---|---|---|---|---|
| 1 | 10 | 5 | 2.314E−04 | 4.6251E+00 | 1.070E−02 | 2.5951E+01 | 6.617E−18 | 2.897E−14 |
| 2 | 15 | 5 | 6.008E+03 | 1.3014E+00 | 1.173E+05 | 7.8666E+00 | 2.877E+00 | 3.654E+01 |
| 3 | 20 | 5 | 4.784E+05 | 5.2542E−01 | 5.028E+06 | 3.4044E+00 | 5.084E+04 | 1.527E+05 |
| 4 | 25 | 5 | 1.943E+06 | 2.6191E−01 | 1.272E+07 | 1.8173E+00 | 1.095E+06 | 1.971E+06 |
| 5 | 30 | 5 | 3.184E+06 | 1.5074E−01 | 1.440E+07 | 1.1186E+00 | 3.183E+06 | 4.569E+06 |
| 6 | 35 | 5 | 3.683E+06 | 9.6114E−02 | 1.239E+07 | 7.6444E−01 | 4.416E+06 | 5.653E+06 |
| 7 | 40 | 5 | 3.656E+06 | 6.7084E−02 | 9.810E+06 | 5.6558E−01 | 4.573E+06 | 5.490E+06 |
| 8 | 45 | 5 | 3.374E+06 | 5.1271E−02 | 7.784E+06 | 4.5042E−01 | 4.239E+06 | 4.903E+06 |
| 9 | 50 | 5 | 3.001E+06 | 4.0313E−02 | 6.048E+06 | 3.6743E−01 | 3.684E+06 | 4.148E+06 |
| 10 | 55 | 5 | 2.608E+06 | 3.4585E−02 | 4.962E+06 | 3.1497E−01 | 3.244E+06 | 3.591E+06 |
| 11 | 60 | 5 | 3.947E+06 | 3.0069E−02 | 7.122E+06 | 2.7788E−01 | 4.895E+06 | 5.355E+06 |
| 12 | 65 | 5 | 2.252E+06 | 2.7723E−02 | 4.058E+06 | 2.5212E−01 | 2.888E+06 | 3.133E+06 |
| 13 | 70 | 5 | 1.603E+06 | 2.5714E−02 | 2.886E+06 | 2.3041E−01 | 2.115E+06 | 2.278E+06 |
| 14 | 75 | 5 | 1.248E+06 | 2.4764E−02 | 2.318E+06 | 2.1537E−01 | 1.733E+06 | 1.858E+06 |
| 15 | 80 | 5 | 1.005E+06 | 2.3908E−02 | 1.923E+06 | 2.0219E−01 | 1.464E+06 | 1.562E+06 |
| 16 | 85 | 5 | 7.765E+05 | 2.3524E−02 | 1.553E+06 | 1.9248E−01 | 1.197E+06 | 1.274E+06 |
| 17 | 90 | 5 | 5.569E+05 | 2.3168E−02 | 1.161E+06 | 1.8375E−01 | 9.061E+05 | 9.616E+05 |
| 18 | 95 | 5 | 3.305E+05 | 2.3182E−02 | 7.280E+05 | 1.7700E−01 | 5.733E+05 | 6.070E+05 |
| 19 | 100 | 5 | 0 | 2.3195E−02 | 0 | 1.7082E−01 | 0 | 0 |
| Sum | | | | | 4.751E+08 | | 2.013E+08 | 2.375E+08 |

For the starting value of $HVL_1 = 0.5$ cm, the ratio of air kermas becomes $K_{air}(0)/K_{air}(HVL_1) = 2.3602$. Iterating the value of $HVL_1$ until we get a ratio of kermas equal to 2 yields a value of $HVL_1 = 0.3803$ cm ($1.0264$ g cm$^{-2}$), the resulting data for the attenuated air kerma after the iteration shown in the rightmost column of the table. Note that

$$K_{air}(0)/K_{air}(0.3803) = 4.751 \times 10^8/2.375 \times 10^8 = 2$$

We have chosen to use Excel/Solver, but any other software would be appropriate; iterating by hand can be laborious.

(b) The corresponding equivalent mass attenuation coefficient in aluminium can be obtained from

$$(\mu/\rho)_{eq} = \frac{\ln 2}{\rho\, HVL_1} = \frac{0.69315}{2.6989 \times 0.3803} = 0.6753 \text{ cm}^2 \text{ g}^{-1}$$

which from the electronic Data Tables for aluminium corresponds to an equivalent energy $k_{eq} = 36.98$ keV. Using this energy to obtain the mass energy-absorption coefficients for water and air from the Data Tables yields

$$[\mu_{en}(k_{eq})/\rho]_w = 0.0842$$

$$[\mu_{en}(k_{eq})/\rho]_{air} = 0.0829$$

$$[\mu_{en}(k_{eq})/\rho]_{w,air} = 1.016$$

(c) Compare with the values obtained in the two previous exercises. If only air kerma-weighted values are considered, the maximum difference in the $\mu_{en}/\rho$-ratio with the spectra averaged value (1.030) is $\sim 1.5\%$. Taking into account the calculations with the different weighting approaches, the largest difference is 2.8% for the energy fluence spectra average (1.044). It can be observed that whereas the iteration procedure to determine the $HVL_1$ is probably the most accurate way to derive the half-value layer, using this $HVL_1$ to derive the equivalent $\mu/\rho$, then derive the equivalent photon energy $k_{eq}$, and thereafter obtain $[\mu_{en}(k_{eq})/\rho]_{med}$ values for water and air results in a rather tortuous path that may lead to poor results.

For completeness, the values of $[\mu_{en}/\rho]_{w,air}$ obtained in Exercises 3–5 using different approaches are summarized in the table below.

| $[\mu_{en}/\rho]_{w,air}$ | Fluence | Energy fluence | Air kerma | Notes |
|---|---|---|---|---|
| From $\bar{k}$ | 1.033 | 1.042 | 1.023 | $\mu_{en}/\rho$ for water and air calculated at the relevant mean photon energy (Exercise 3) |
| Spectra averaged | 1.036 | 1.044 | 1.030 | For example, see Exercise 4 $$[(\bar{\mu}/\rho)_{w,air}]_\Phi = \frac{\sum[\mu(k)/\rho]_{w,air}\Phi_k \Delta k}{\sum \Phi_k \Delta k}$$ |

| $[\mu_{en}/\rho]_{w,air}$ | Fluence | Energy fluence | Air kerma | Notes |
|---|---|---|---|---|
| As ratios of means | 1.024 | 1.030 | 1.020 | For example, see Exercise 4 $$\frac{[(\bar{\mu}/\rho)_w]_\Phi}{[(\bar{\mu}/\rho)_{air}]_\Phi} = \frac{\sum[\mu(k)/\rho]_w\Phi_k\Delta k}{\sum[\mu(k)/\rho]_{air}\Phi_k\Delta k}$$ |
| From HVL$_1$ in Al | | | 1.016 | Exercise 5, see text |

**6** Assuming that for megavoltage (high-energy) photon beams the effective attenuation coefficient in water at a constant source-to-detector distance (SDD = 100 cm) can be approximated by the linear attenuation coefficient of the spectrum mean energy ($\bar{k} \approx$ MV/3), estimate the TPR$_{20,10}$ and the percentage depth dose at 10 cm depth, PDD(10), at an SSD of 100 cm, for a 25 MV beam.
*Answer: TPR$_{20,10}$ = 0.788; PDD(10,100)=83.7*

**Solution:**
The effective attenuation coefficient at a constant SDD corresponds to the parameter $\bar{\mu}'$ in the textbook. For the PDD(10) calculation we use the modified Greening expression

$$D_w(z) \approx \Psi_0 \, \frac{\mu_{en} \, \bar{\mu}_e}{\bar{\mu}_e - \bar{\mu}_p} [e^{-\bar{\mu}_p z} - e^{-\bar{\mu}_e z}]$$

for which we need to determine the photon and electron effective attenuation coefficients at an SSD = 100 cm, $\bar{\mu}_p$ and $\bar{\mu}_e$, respectively. Note that, as we calculate relative depth doses, the terms in front of the square brackets, constant for all depths, can be omitted.

The approximation $\bar{\mu}' \approx \mu(\bar{k})$ for high-energy MV beams is justified in terms of the plot in Figure 7.2 for the various MV attenuation coefficients in Table 7.2 of the textbook (adopted from BJR-25) and the photon linear attenuation coefficients $\mu$, $\mu_{tr}$, and $\mu_{en}$ for water evaluated at the relevant mean photon energy $\bar{k}$.

Hence, for 25 MV, $\bar{k} \approx 25/3 = 8.33$ MeV, and from the electronic Data Tables $\bar{\mu}' \approx \mu(\bar{k}) = 0.0238$ cm$^{-1}$. We get TPR$_{20,10}$ using Eq. (7.15) in the textbook

$$\bar{\mu}' = 0.1 \times \ln(1/\text{TPR}_{20,10})$$

that is,

$$\text{TPR}_{20,10} = e^{-10 \, \bar{\mu}'} = 0.788$$

and $\bar{\mu}_p$ from Eq. (7.16)

$$\bar{\mu}_p \approx 0.0174 + \bar{\mu}' = 0.0412 \text{ cm}^{-1}$$

The electron effective attenuation coefficient is approximated by $\bar{\mu}_e = 1/R_{CSDA}$, where the continuous slowing-down range is to be evaluated at the mean energy of the equilibrium slowing-down spectrum of the

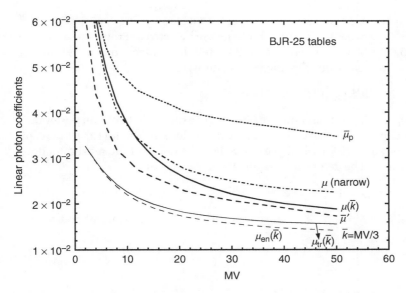

**Figure 7.2** Different photon coefficients in water as a function of the nominal acceleration potential.

secondary electrons produced by the MV photons (see Section 9.3.2 in the textbook). Assuming that all the secondary electrons are liberated through Compton interactions, the mean initial electron energy is given by

$$\bar{E}_0 = \bar{k}\, \frac{\sigma_{\mathrm{tr}}^{\mathrm{KN}}(\bar{k})}{\sigma^{\mathrm{KN}}(\bar{k})}$$

where $\sigma^{\mathrm{KN}}$ and $\sigma_{\mathrm{tr}}^{\mathrm{KN}}$ are the total and energy-transfer Klein–Nishina cross sections. The mean energy of the equilibrium slowing-down spectrum of the secondary electrons is approximately given by $\bar{E}_z \sim 0.5\bar{E}_0$.

The Klein–Nishina cross sections for free electrons can be obtained from the Data Tables. As a numerical illustration, we use instead the full expressions given in Chapter 3, that is, Eqs (3.84) and (3.85) (see also Fig. 3.18):

$$\sigma_{\mathrm{C,KN}}(\epsilon) = 2\pi r_{\mathrm{e}}^2 \left\{ \frac{1+\epsilon}{\epsilon^2} \left[ \frac{2(1+\epsilon)}{1+2\epsilon} - \frac{\ln(1+2\epsilon)}{\epsilon} \right] \right.$$

$$\left. + \frac{\ln(1+2\epsilon)}{2\epsilon} - \frac{1+3\epsilon}{(1+2\epsilon)^2} \right\}$$

and

$$\sigma_{\mathrm{C,KN}}^{\mathrm{tr}} = 2\pi r_{\mathrm{e}}^2 \left\{ \frac{2(1+\epsilon)^2}{\epsilon^2(1+2\epsilon)} - \frac{1+3\epsilon}{(1+2\epsilon)^2} - \frac{(1+\epsilon)(2\epsilon^2-2\epsilon-1)}{\epsilon^2(1+2\epsilon)^2} \right.$$

$$\left. - \frac{4\epsilon^2}{3(1+2\epsilon)^3} - \left[ \frac{1+\epsilon}{\epsilon^3} - \frac{1}{2\epsilon} + \frac{1}{2\epsilon^3} \right] \ln(1+2\epsilon) \right\}$$

where $\epsilon = \bar{k}/m_{\mathrm{e}}c^2$, $m_{\mathrm{e}}c^2 = 0.511$ MeV, and $r_{\mathrm{e}} = 2.81794 \times 10^{-13}$ cm.

For $\bar{k} = 8.33$ MeV, we get $\epsilon = 16.31$, yielding $\sigma^{\mathrm{KN}}(\bar{k}) = 5.8165 \times 10^{-26}$ and $\sigma^{\mathrm{KN}}_{\mathrm{tr}}(\bar{k}) = 3.8988 \times 10^{-26}$, both in cm²/electron. From these we get $\bar{E}_0 = 5.6$ MeV and therefore $\bar{E}_z = 2.8$ MeV, for which from the electronic Data Tables (electron stopping powers for water)

$$\bar{\mu}_e = 1/R_{\mathrm{CSDA}}(2.8) = 0.71 \ \mathrm{cm}^{-1}$$

With $\bar{\mu}_p$ and $\bar{\mu}_e$, we can now use Greening's expression to obtain dose values at any depth. These have to be relative to the dose maximum (to get percentage depth doses), for which we derive first the depth of the maximum dose, $z_{\max}$. This can be determined taking the derivative of

$$D(z_{\max}) = e^{-\bar{\mu}_p z_{\max}} - e^{-\bar{\mu}_e z_{\max}}$$

that is, $\bar{\mu}_e\, e^{-\bar{\mu}_e z_{\max}} - \bar{\mu}_p\, e^{-\bar{\mu}_p z_{\max}}$, setting it to zero (for maximum) and solving for $z_{\max}$,

$$z_{\max} = \frac{\ln(\bar{\mu}_e/\bar{\mu}_p)}{\bar{\mu}_e - \bar{\mu}_p}$$

This yields $z_{\max} = 4.25$ cm and therefore $D(z_{\max}) = 0.79$. Together with $D(10) = 0.66$, we get

$$\mathrm{PDD}(10) = 100\,\frac{D(10)}{D(z_{\max})} = 83.7$$

or the entire relative depth-dose distribution at SSD = 100 cm illustrated in Figure 7.3

It can be seen that the distribution does not show a contribution from electron contamination (note that $D(0) \to 0$), as this is only an approximation to the distribution of a real 25 MV beam.

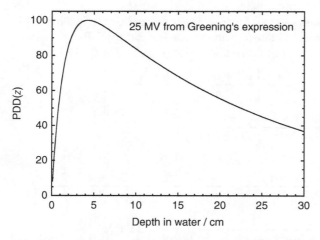

**Figure 7.3** Percent depth-dose distribution in water of 25 MV photons, calculated using Greening's expression.

To verify the self-consistency of the expressions used, we can derive $TPR_{20,10}$ from the depth doses at $SSD = 100$ cm using the approximation of Eq. (7.12):

$$D^{SDD}(z) \approx D^{SSD}(z)\left(\frac{SSD + z}{SSD}\right)^2,$$

which gives $D^{SDD}(10) = 0.80$ and $D^{SDD}(20) = 0.63$, their ratio being $TPR_{20,10} = 0.789$, that is, in good agreement with the value obtained above. Compared with the average values in Table 7.2 of the textbook for realistic 25 MV beams, $z_{max} = 3.8$ cm, $PDD(10) = 83.0$, and $TPR_{20,10} = 0.799$, the differences are within approximately 1% (except for $z_{max}$), which can be justified in terms of the different accelerators used to arrive at the average values given in the table.

**7** A radiation therapy 250 MeV proton beam has an experimentally-determined practical range in water of 38.12 cm and the middle of the spread-out Bragg peak (SOBP) falls at a depth of 34 cm. Estimate the most probable energy of the protons at the middle of the SOBP.
*Answer: 70.2 MeV*

**Solution:**
The goal of this exercise is to illustrate the meaning of the process underlying the use of the residual range $R_{res}$ as a parameter to specify the quality of a proton beam. This is defined as

$$R_{res} = R_p - z_{ref}$$

where $R_p$ is the practical range and $z_{ref}$ the reference depth for beam calibration, taken to be at the middle of the SOBP.

Figure 7.4 shows the relation between proton energy and CSDA range in water, $R_{CSDA}$, which is very close to $R_p$.

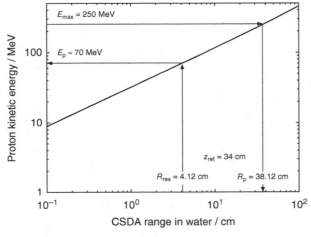

**Figure 7.4** Relation between proton kinetic energy and CSDA range in water, $R_{CSDA}$, which is very close to $R_p$.

When $z_{ref} = 34$ cm is subtracted from $R_p$ yielding $R_{res} = 4.12$ cm, the arrows in the plot indicate that this corresponds to the $R_{CSDA}$ of a proton having 70.2 MeV (interpolating from the electronic Data Tables), which approximates the value of the most probable energy at $z_{ref}$ (the middle of the SOBP). Hence, $R_{res}$ is an indirect estimate of the most probable proton energy at $z_{ref}$.

**8** Repeat the previous exercise for a 200 MeV proton having a practical range of 26.1 cm and the middle of the SOBP falls at a depth of 22 cm. Compare with the result above and draw conclusions.
*Answer: 69.9 MeV*

**Solution:**
It was emphasized in the textbook that in contrast to the beam quality specifiers for other radiation types, for example, $R_{50}$ for electron beams, $R_{res}$ is not unique to a particular beam but will be the same for the same distance $R_p - z_{ref}$ in any proton beam, independently of $R_p$.
In this case we have a rather different incident beam and $R_p$, 200 MeV and ~26 cm as opposed to 250 MeV and ~38 cm in the previous exercise, that is, their maximum penetration differs considerably. However, in this case

$$R_{res} = R_p - z_{ref} = 26.1 - 22 = 4.1 \text{ cm}$$

that is, it practically coincides with the previous case, corresponding to a most probable proton energy of 69.9 MeV at $z_{ref}$.
The two beams have therefore the same beam quality index because by definition this is a characteristic of the beam properties at $z_{ref}$ (the middle of the SOBP), irrespective of the energy and penetration properties of the incident beam. Consequently their dosimetric parameters will be identical in terms of this index as both have the same most probable energy at $z_{ref}$.

# 8

# The Monte Carlo Simulation of the Transport of Radiation through Matter

1   Calculate the value of $\pi$ by generating random points on a square and counting the proportion that lie inside an inscribed circle.

**Solution:**

This is the simplest Monte Carlo estimate for the value of $\pi$. The probability of a point landing in the circle is proportional to the relative areas of the circle and the square, that is,

$$P = \frac{\pi R^2}{L^2} = \frac{\pi R^2}{(2R)^2} = \frac{\pi}{4},$$

which can be written as $\pi = 4\,P$, where the probability $P$ is equal to the number of points within the circle ($n_{hits}$) divided by the total number of points within the square ($n_{trials}$).

With reference to Figure 8.1, and using any uniform random number generator producing numbers in the interval [0,1], a minimal programming code can be written where $(x, y)$ coordinates are generated within the interval $[-L/2, +L/2]$, and a check is made of those within the circle:

$n_{hits} = 0$

loop for $n = 1, n_{trials}$

$\quad x = -L/2 + L \times \mathrm{ran}[0, 1]$

$\quad y = -L/2 + L \times \mathrm{ran}[0, 1]$

$\quad$ if $(x^2 + y^2 \leq R)$ $n_{hits} = n_{hits} + 1$

end loop

and then $\pi = 4\,P = 4\,\frac{n_{hits}}{n_{trials}}$.

One needs at least $10^5$ trials to achieve convergence to the true value of $\pi$ within 2 or 3 places, that is, the Monte Carlo estimate of $\pi$ is rather inefficient. Note that the $(x, y)$ points could equally well have been generated within the first quadrant $[0, L/2]$. The code above is intended to illustrate the generation of random numbers within an interval $[a, b]$, with $a < b$, that is, $\xi = a + (b - a) \times \mathrm{ran}[0, 1]$.

*Fundamentals of Ionizing Radiation Dosimetry: Solutions to Exercises,* First Edition.
Pedro Andreo, David T. Burns, Alan E. Nahum, and Jan Seuntjens.
© 2017 Wiley-VCH Verlag GmbH & Co. KGaA. Published 2017 by Wiley-VCH Verlag GmbH & Co. KGaA.

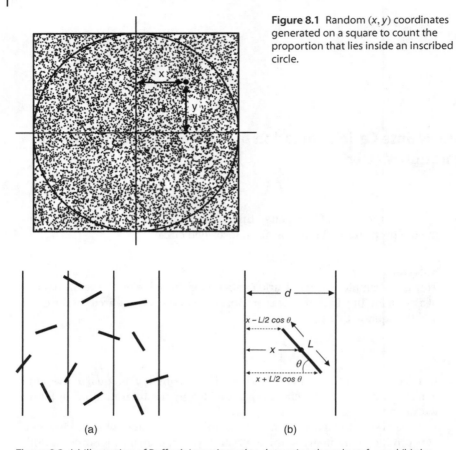

**Figure 8.1** Random $(x, y)$ coordinates generated on a square to count the proportion that lies inside an inscribed circle.

**Figure 8.2** (a) Illustration of Buffon's 'experiment' to determine the value of $\pi$ and (b) the geometry of a needle relative to two parallel lines.

**2** Simulate Buffon's needle problem to estimate the value of $\pi$ by throwing needles of length $L$ at random onto a horizontal plane ruled with parallel straight lines at distance $d$ ($d > L$) apart. Determine first the analytical expression for the probability of a needle intersecting a line, $P = 2\,L/\pi d$.

**Solution:**
Figure 8.2 illustrates Buffon's 'experiment' (a) and the geometry of a needle relative to two parallel lines (b). There are two random variables to consider, the position $x$ of the needle's center, which will be uniformly distributed in the interval $[0, d/2]$, and the needle's orientation $\theta$ in the interval $[0, \pi/2]$. An intersection will occur when the needle's center position is smaller than or equal to the $x$-projection of the needle's half, that is, $x \leq (L/2) \cos \theta$.

To determine the expression for the probability of an intersection, recall first that the probability density function of a uniform distribution of a variable $y$ within the interval $[y_{\min}, y_{\max}]$ is $1/(y_{\max} - y_{\min})$ and 0 elsewhere. According to this

$$f(x \in [0, d/2]) = \frac{2}{d} \quad \text{and} \quad f(x \in [0, \pi/2]) = \frac{2}{\pi}$$

and as the two variables are uncorrelated, the combined probability density function is the product of the two, that is,

$$f(x \in [0, d/2], \theta \in [0, \pi/2]) = \frac{4}{d\pi}$$

The probability of an intersection is then

$$P = \int_0^{\pi/2} \int_0^{(L/2)\cos\theta} \frac{4}{d\pi} \, d\theta \, dx = \frac{2L}{d\pi}$$

and from this expression $\pi = \frac{2L}{dP}$.

In a Monte Carlo simulation, the probability $P$ is equal to the number of intersections ($n_{\text{hits}}$) divided by the total number of trials ($n_{\text{trials}}$). The code to perform this simulation hence is

$n_{\text{hits}} = 0$

loop for $n = 1, n_{\text{trials}}$

$\quad x = \text{ran}[0, 1] \times \dfrac{d}{2}$

$\quad \theta = \text{ran}[0, 1] \times \dfrac{\pi}{2}$

$\quad$ if $(x \leq \dfrac{L}{2} \cos\theta)$ $n_{\text{hits}} = n_{\text{hits}} + 1$

end loop

and then $\pi = \dfrac{2L}{d\, \frac{n_{\text{hits}}}{n_{\text{trials}}}}$.

One needs at least $10^5$ trials to achieve convergence to the true value of $\pi$ within 2 or 3 places, that is, the Monte Carlo estimate of $\pi$ is rather inefficient. A graphical description of this exercise can be found in, for example, http://en.wikipedia.org/wiki/Buffon's-needle.

3  Using a Monte Carlo simulation, (a) solve the triple integral of the function $f(x, y, z) = g(x)\, g(y)\, g(z)$, where $g(v) = v e^{-v^2/2}$, between the limits $a = 0$ and $b = 4$ for each integral, that is,

$$\int_a^b \int_a^b \int_a^b f(x, y, z) \, dx \, dy \, dz$$

(b) Calculate the relative standard uncertainty (in percent) of the result. *Answer: 0.998, with relative standard uncertainty ±0.2% for the present case using $10^6$ trials*

**Solution:**

The function $f(x, y, z) = g(x)\, g(y)\, g(z)$ is given by $f(x, y, z) = x\, y\, z\, e^{-(x^2+y^2+z^2)/2}$ and the integral to be solved is

$$\int_0^4 \int_0^4 \int_0^4 x\, y\, z\, e^{-(x^2+y^2+z^2)/2} \; dx\, dy\, dz$$

Equation (8.29) in the textbook (written there for one dimension) results from the *law of large numbers*, stating that the average of the results obtained from a large number of trials should be close to the expected mean value and becomes closer as more trials are performed; this is the most common method for solving high-dimensional integrals. For three dimensions it can be written as

$$\langle I \rangle = (b - a)^3 \frac{1}{N} \sum_1^N f(x_i, y_i, z_i),$$

which corresponds to the volume of the integrating space enclosure multiplied by the mean value of $f(x, y, z)$ after $N$ trials, that is, $\langle I \rangle = V \times \bar{f}$. The mean value $\bar{f}$ will be determined by random sampling, with the value of the function being calculated for each trial of the triplet $(x_i, y_i, z_i)$ sampled between the limits $a$ and $b$.

The code to perform this simulation is, for $N$ trials,

$a = 0$; $b = 4$

$V = (b - a)^3 = 64$

sumof $= 0$; sumof2 $= 0$

loop for $n = 1, N$

    $x = a + (b - a) \times \text{ran}[0, 1]$

    $y = a + (b - a) \times \text{ran}[0, 1]$

    $z = a + (b - a) \times \text{ran}[0, 1]$

    sumof $=$ sumof $+ f(x, y, z)$

    sumof2 $=$ sumof2 $+ f(x, y, z)^2$

end loop

From a loop with $10^6$ trials, we get sumof $= 15\,600.8$ and sumof2 $= 1358.58$. Hence

$$\bar{f} = \frac{\sum_{i=1}^N f(x_i, y_i, z_i)}{N} = \frac{\text{sumof}}{N} = \frac{15600.8}{10^6} = 0.01560$$

that is, $\langle I \rangle = V \times \bar{f} = 64 \times 0.0156008 = 0.99845$. This can be compared with a numerical quadrature, which gives $0.99899$.

For the standard uncertainty of $\langle I \rangle$, we use Eq. (8.35), multiplied by the volume $V$, that is,

$$s_{\langle I \rangle} = V \sqrt{\frac{1}{N(N-1)} \left[ \sum_{i=1}^N x_i^2 - \frac{\left( \sum_{i=1}^N x_i \right)^2}{N} \right]}$$

$$= V \sqrt{\frac{1}{N(N-1)} \left[ \text{sumof2} - \frac{(\text{sumof})^2}{N} \right]}$$

$$= 64 \sqrt{\frac{1}{10^6(10^6-1)} \left[ 1358.58 - \frac{15600.8^2}{10^6} \right]} = 0.00214,$$

which as a percentage of $\langle I \rangle$ yields a relative standard uncertainty of 0.21%. Other examples of MC integrals can be found in, for example, http://mathfaculty.fullerton.edu/mathews/n2003/montecarlopimod.html.

**4** A 6 MeV electron beam incident on a 0.1 mm foil of gold has a Gaussian-like radial profile (centered on the beam axis) at the exit of the foil with an FWHM of 1 mm. (a) Determine the variance of the distribution and (b) perform a Monte Carlo sampling of the transverse coordinates, making a histogram of the sampled values. As verification, compare with the corresponding analytically determined Gaussian distribution.
*Answer: $\sigma_{Gaussian} = 0.425$ mm (FWHM = 2.3548 $\sigma$). Sampling is made using Eq. (8.19) in the textbook.*

**Solution:**

(a) The first step is to derive the relation between the FWHM and the variance of a Gaussian distribution. The distribution is given by

$$g(r) = \frac{1}{\sqrt{2\pi\sigma^2}} \, e^{-(r-\bar{r})^2/2\sigma^2}$$

The extreme points $r_-$ and $r_+$ at half-maximum are, by definition, at a value $g(r)/2$. Hence solving for $r$ the equation

$$e^{-(r-\bar{r})^2/2\sigma^2} = 0.5$$

(taking the logarithm of both sides) results in

$$r_- = \bar{r} - 1.1774 \, \sigma$$
$$r_+ = \bar{r} + 1.1774 \, \sigma$$

and the full width at half maximum is

$$\text{FWHM} = r_+ - r_- = 2.3548 \, \sigma$$

which, for FWHM = 1 mm, results in $\sigma = 0.4247$ mm.

(b) As the Gaussian is centered on the axis, $\bar{r} = 0$, and therefore calculating values of a Gaussian distribution with $\sigma = 0.4247$ is straightforward. For convenience, we normalize the values at the maximum value (at $r = 0$), which is 0.9394. The distribution is plotted in Figure 8.3.
We then proceed with the sampling from a Gaussian using Eq. (8.19), and to improve the MC statistics, we only consider positive values.
In order to score the sampled $r$-values and build a histogram of, say, 200 bins, we choose a maximum distance $r_{max} = 2 \times \text{FWHM} = 2$ mm; this means that our bin width will be $\Delta r = r_{max}/200 = 0.01$ mm.

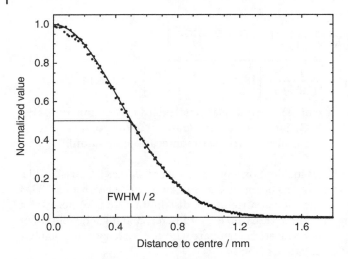

**Figure 8.3** Random sampling from a Gaussian distribution (dots). The solid line corresponds to the analytically determined Gaussian.

Using the Box–Muller algorithm to derive a Gaussian random number distribution from a uniform distribution, the code to perform the simulation for $N$ trials is

$\text{histo}(i) = 0$            [reset the histogram bins]

loop for $n = 1, N$

    $R_1 = \text{ran}[0, 1]$

    $R_2 = \text{ran}[0, 1]$            [select two random numbers]

    $y_1 = \sqrt{-2 \ln R_1} \, \cos(2\pi R_2)$            [Eq . (8.19)]

    $y_2 = \sqrt{-2 \ln R_1} \, \sin(2\pi R_2)$

    $i = \dfrac{|y_1|}{\Delta r} + 1$        [index $i$ for histo; must be an integer number]

    $\text{histo}(i) = \text{histo}(i) + 1$

    repeat for $y_2$

end loop

Note that the method yields 2 $N$ random numbers per $N$ trials.

Note also that depending on the random number generator used, ln $R$ might become undefined if $R = 0$. This can be avoided using $(1 - R)$ instead of $R$. Similar precautions are needed if the generator yields $R = 1$. Most generators avoid both extreme values for these reasons.

Once the loop is completed, search for the maximum value of the histogram and normalize all the values to it. For $10^5$ trials, our sampled values are plotted in Figure 8.3 (dots), which also includes the analytically determined values (line). It can be observed that the agreement is excellent, preserving the width of the Gaussian (the figure shows FWHM/2).

**5** (a) Calculate by Monte Carlo sampling a histogram of the polar angular distribution of the electrons in the previous exercise (6 MeV incident on a 0.1 mm gold foil), assuming that the distribution is Gaussian. Compare with the analytically determined Gaussian distribution. (b) Derive the relation between the FWHM and $\sigma$. (See Chapter 2 in the textbook; mass scattering powers, $T/\rho$, are given in the electronic Data Tables.)
*Answer: (a) Using $T/\rho$, $\sigma = 0.5$ rad; the CPD yields samples $\theta = \sigma\sqrt{-\ln \xi}$.*
*(b) FWHM $= 1.6651\,\sigma$.*

**Solution:**

(a) Recall that under the Gaussian approach for multiple small-angle scattering (see Chapter 2), the polar angular distribution is derived as the *product* of two symmetrical Gaussians for the projected angles $\theta_x$ and $\theta_y$, yielding[1]

$$f(\theta) = \frac{1}{\pi\,\sigma^2}\,\exp\left(-\frac{\theta^2}{\sigma^2}\right)$$

that is, the polar angle distribution is not given by a single Gaussian, and the sampling from this function needs to be modified accordingly. Its CPD is

$$F(\theta) \equiv \xi = \int_0^\theta f(\theta)\,2\pi\theta\,d\theta = 1 - \exp\left(-\frac{\theta^2}{\sigma^2}\right)$$

from where

$$\theta = \sigma\sqrt{-\ln\,(1-\xi)}$$

and we can use the Box–Muller method (Eq. (8.19) in the textbook) to sample two angles using two random numbers in each iteration (see below).
From the Data Tables, the mean square scattering angle for 6 MeV electrons in gold ($\rho = 19.32$ g cm$^{-3}$) is $T/\rho = 1.2947$ rad$^2$ cm$^2$ g$^{-1}$. Hence, for a 0.1 mm thickness $t = 0.01 \times 19.32 = 0.1932$ g cm$^{-2}$,

$$\sigma = \sqrt{T/\rho \times t} = \sqrt{1.2947 \times 0.1932} = 0.5001 \text{ rad } (28.66°)$$

To score the sampled $\theta$-values and build a histogram of, say, 200 bins, we choose a maximum $\theta_{\max} = 4\,\sigma$ rad; this means that our bin width will be $\Delta\theta = \theta_{\max}/200 \approx 0.01$ rad. The code to perform the simulation for $N$ trials is

histo($j$) = 0                                                 [reset the histogram bins]

loop for $n = 1, N$

    $R_1 = $ ran[0, 1]; $R_2 = $ ran[0, 1]               [select two random numbers]

    $\theta_1 = \sigma\sqrt{-\ln R_1}\,\cos(2\pi R_2)$              [Eq. (8.19) modified, see above]

---

1 Recall that for small angles $\bar{\theta}^2 \approx \bar{\theta}_x^2 + \bar{\theta}_y^2$, and because the distributions are symmetrical, $\bar{\theta}^2 = 2\bar{\theta}_x^2$; likewise, $\theta^2 \approx \theta_x^2 + \theta_y^2$.

$$\theta_2 = \sigma\sqrt{-\ln R_1}\,\sin(2\pi R_2)$$

$$j = \frac{|\theta_1|}{\Delta\theta} + 1 \qquad\qquad \text{[index } j \text{ for histo; must be an integer number]}$$

$$\text{histo}(j) = \text{histo}(j) + 1$$

repeat for $\theta_2$

end loop

Note that the method yields $2\,N$ random numbers per $N$ trials.

Once the loop is completed, search for the maximum value of the histogram and normalize all the values to it. For $10^5$ trials, our sampled values are plotted in Figure 8.4 (dots), which also includes the analytically determined values (solid line). For comparison, the analytically derived values for a single Gaussian having the same variance, that is,

$$g(\theta) = \frac{1}{\sqrt{2\pi\sigma^2}}\, e^{-\theta^2/2\sigma^2}$$

are also included in the figure (dashed line), illustrating the error that would occur if such an incorrect function had been used.

(b) As in the previous exercise, we need now solve for $\theta$ the equation

$$e^{-\theta^2/\sigma^2} = 0.5$$

Taking the logarithm of both sides results in

$$\theta_- = -0.8326\,\sigma$$

$$\theta_+ = +0.8326\,\sigma$$

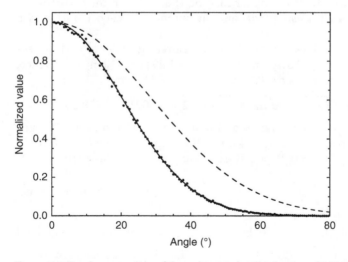

**Figure 8.4** Random sampling of the polar angular distribution of 6 MeV electrons incident on a 0.1 mm gold foil, assuming that the distribution is Gaussian (dots). The solid line corresponds to the analytically determined Gaussian. The dashed line corresponds to analytically derived values for a single Gaussian having the same variance.

and the full width at half maximum is

$$\text{FWHM} = \theta_+ - \theta_- = 1.6651\,\sigma$$

which, for $\sigma \approx 0.5$ rad, results in $\text{FWHM} = 0.83$ rad ($47.7°$) and $\text{FWHM}/2$ $\approx 24°$ as can be seen in the figure.

6 Using any of the Monte Carlo systems available, calculate the absorbed dose profile in the incident direction of a 1 MeV photon broad beam impinging on a 20 mm–10 mm–20 mm three-slab geometry using a low electron transport energy cut-off $E_{cut}$ (e.g., 1 keV or 5 keV) and a high $E_{cut}$ (e.g., 1 MeV), with materials (a) water–aluminium–water and (b) aluminium–water–aluminium. Discuss the reasons for the dose patterns at the different interfaces. (Note that to visualize interface effects, a high resolution, <1 mm, is required near the interfaces.)

**Solution:**
*Readers should be aware that for energies and material thicknesses such that radiative losses are not significant, D(high $E_{cut}$) is commonly used to estimate the kerma, noting that when bremsstrahlung production is significant, this method will underestimate the kerma by neglecting the increasing contribution of bremsstrahlung with depth.*

For this exercise we have relied on the user code *DOSRZ* of the *EGSnrc* MC system, with a very large radius (100 cm) and a broad beam configuration (source number 2). This makes the beam size as broad as the scoring geometry, so that lateral scattering does not play a role. Our resolution on both sides of the interfaces is 0.025 cm and 0.1 cm elsewhere, and we have added a thick slab of the last material at the end of the three slabs (to avoid having vacuum). Default transport parameters are used except for the electron energy cut-offs, and the *PEGS4* data files are identical in all cases ($\Delta = \text{AE} = \text{AP} = 1$ keV). Figures 8.5 and 8.6 show the results obtained for the material configurations (a) and (b), where $E_{cut} = 1$ keV (solid lines) or 1 MeV (dashed lines). The 1 MeV photon incident direction is from left to right.

As emphasized in Chapter 2 of the textbook, the minimum energy at which stopping powers of light charged particles should be calculated is several times the $K$-shell ionization threshold, as the theory does not include shell corrections. For aluminium $U_K = 1.56$ keV, and therefore the accuracy of a calculation below about 5 keV or 10 keV is questionable; the exercise thus does not aim at being rigorous in this sense.

The first observation to be made in both plots is that away from the interfaces or the entrance surface, the absorbed dose determined with the two electron energy cut-offs are practically identical. This is expected, as the relatively thick slabs of each material are homogeneous and without boundaries between the scoring sub-slabs.

A completely different pattern occurs at the interfaces and serves as an illustration of the importance of electron transport in photon calculations and how MC algorithms deal with boundaries, more or less independently of the cross-section data used.

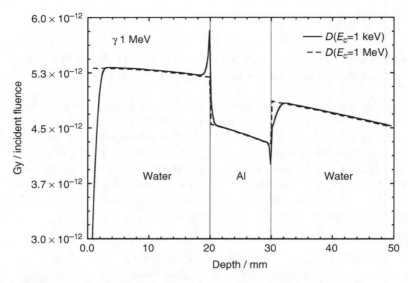

**Figure 8.5** Absorbed dose profile in the incident direction of a 1 MeV photon broad beam impinging on a three-slab geometry water–aluminium–water using electron transport cut-off energies of 1 keV and 1 MeV.

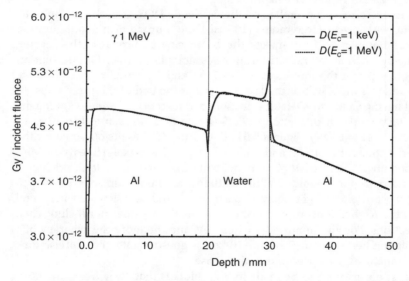

**Figure 8.6** Absorbed dose profile in the incident direction of a 1 MeV photon broad beam impinging on a three-slab geometry aluminium–water–aluminium using electron transport cut-off energies of 1 keV and 1 MeV.

(a) For the solid curve in Figure 8.5, with $E_{cut} = 1$ keV, there is first a typical electron build-up at the entrance surface due to the transport of the photon-generated electrons in the first millimeter or so. This build-up is not present in the $E_{cut} = 1$ MeV case (dashed curve), as any electron produced is stopped and absorbed at its site of production ('on the spot').

When the particles in water approach the interface with aluminium, of higher atomic number, there is considerable electron backscatter from it that raises dramatically the dose prior to the boundary. From that interface onward, the dose decreases smoothly until its equilibrium value in aluminium, smaller than in water due to the difference in stopping powers and larger mass; the dose inside aluminium follows an approximate exponential pattern with slope different from that in water for the same reasons. Note that in this case no build-up is seen, as this is more than offset by electrons entering from the water.

At the next interface with water, the opposite effect takes place. Electron backscatter from water cannot balance the electron fluence from aluminium, so that the dose decreases. From this interface onward the dose pattern is again that of electron build-up, as at the entrance surface, although reduced in size because some electrons enter from the aluminium. None of the effects mentioned can be observed in the case of the high electron energy cut-off, $E_{cut} = 1$ MeV, as electrons are not allowed to travel (they are absorbed on the spot).

(b) For the configuration in Figure 8.6, the trend at the interfaces (except at the entrance surface) is reversed, that is, a decrease of the dose going from aluminium to water, higher dose in water following a build-up, large electron backscatter from aluminium. The arguments for the dose increase and decrease are identical to case (a).

Note that if behind the last slab we would have considered vacuum, in both cases there would have been lack of backscatter and the dose in the last sub-slabs would have decreased.

7 Calculate (with any Monte Carlo system) the fluence spectrum differential in energy of a 2.5 MeV electron broad beam incident on a 2 mm thick layer of silicon. Consider the cases where both the electron transport energy cut-off $E_{cut}$ and the threshold $\Delta$ for knock-on electrons are equal to 1 keV and to 10 keV (i.e., $E_{cut} = \Delta = 1$ keV and $E_{cut} = \Delta = 10$ keV) and discuss the reasons for the possible differences in the overlapping energy regions.

**Solution:**

For comparison purposes we have used two MC systems in this exercise. The user codes were *EGSnrc/FLURZ* and *Penelope/PenEasy*, both simulating the transport of a broad electron beam incident on a 2 mm thick slab of silicon. Default transport parameters are used except for the electron energy cut-offs, and data files have been generated for the two systems (*EGSnrc/FLURZ* requires a different dataset for each $\Delta$, while only one is required for *Penelope/PenEasy*). In both cases we have selected 1000 logarithmic intervals, but such large number is unnecessary unless some of the details below need to be shown.

As emphasized in Chapter 2 of the textbook, the minimum energy at which stopping powers of light charged particles should be calculated is several times the K-shell ionization threshold as the theory does not include shell corrections. For silicon $U_K = 1.84$ keV, and therefore the accuracy of the

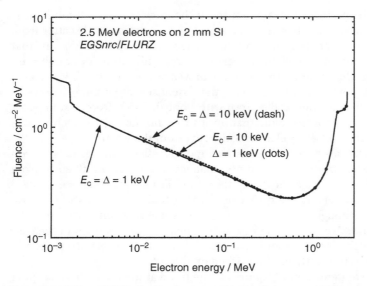

**Figure 8.7** *EGSnrc/FLURZ* calculated spectra of 2.5 MeV electrons incident on a 2 mm thick layer of silicon using different energy transport cut-offs ($E_c$) and threshold for the production of knock-on electrons ($\Delta$).

calculations below about 5 keV or 10 keV is questionable; the exercise thus does not aim at being rigorous in this sense.

(a) The results for the *EGSnrc/FLURZ* calculation of the spectrum are shown in Figure 8.7. This code outputs fluence in units of 'MeV$^{-1}$ per incident fluence', that is, cm$^{-2}$ MeV$^{-1}$.

If the remark on stopping powers above is not taken into account, in addition to the difference in cut-off and threshold energies, a small increasing difference can be observed below a few hundred keV between the solid and dashed curves. The reason for this discrepancy is expected to be due to the different $\Delta$-values used in the restricted stopping powers, as energy straggling below $\Delta$ is neglected. Recall that in class II electron transport schemes, energy straggling between $\Delta$ and the maximum energy loss is accounted for during the sampling of major or hard inelastic collisions, but unless an additional energy loss sampling is made below $\Delta$, straggling below this threshold is ignored.

To verify this assumption, a calculation has been made using $E_{cut} = 10$ keV and $\Delta = 1$ keV, on the basis that even if electrons generated between 1 keV and 10 keV will not be transported, this choice ensures that energy-loss straggling down to 1 keV is taken into account. The resulting spectrum is included in Figure 8.7 as dots, where now excellent agreement with the $E_c = 1$ keV and $\Delta = 1$ keV can be observed.

(b) The results for the *Penelope/PenEasy* calculation of the spectrum (called 'tally-Fluence-TrackLength-electron.dat') are shown in Figure 8.8. In this code the differential fluence is integrated over the detector volume; hence its units are cm$^3$/(cm$^2$ eV) = cm/eV per incident particle. In

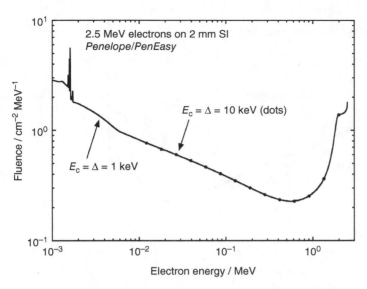

**Figure 8.8** *Penelope/PenEasy* calculated spectra of 2.5 MeV electrons incident on a 2 mm thick layer of silicon using different energy transport cut-offs ($E_c$) and threshold for the production of knock-on electrons ($\Delta$).

addition to the $10^6$ factor between eV and MeV, to convert into $cm^{-2}$ $MeV^{-1}$, we divide the output by the slab thickness, 0.2 cm, and take it to be per incident fluence.

The first observation to be made is that in this case no significant difference between the two $E_{cut}$ and $\Delta$ calculations can be observed. This is explained in terms of the Gaussian energy sampling below $\Delta$ made in *Penelope*, which, even if approximate, provides consistency between the two simulations. The second comment is on the several peaks that appear just below the K-shell edge, produced by electron impact ionization (we have unsuccessfully tried to reproduce these with *FLURZ*, even with the option of using the *Penelope* electron impact database). In addition *Penelope* includes shell corrections to the stopping powers, at least in an approximate manner, due to its use of the GOS rather than tabulated data. These are all minor details that usually do not affect dosimetry calculations, but might be of interest for low energy simulations.

The $E_{cut} = \Delta = 1$ keV spectra calculated with the two MC systems are shown in Figure 8.9. Here, in addition to the already indicated difference at very low energies, it can be seen that above 10 keV the difference between the two spectra is similar to that shown in Figure 8.7 for the two initial calculations with *EGSnrc/FLURZ*. This difference is not expected to substantially affect dosimetry calculations, but might be partially responsible for the small discrepancies (within about 0.5%) indicated in the scientific literature for dose calculations with the two MC systems; see also the two values of $D(MC)$ in the next exercise.

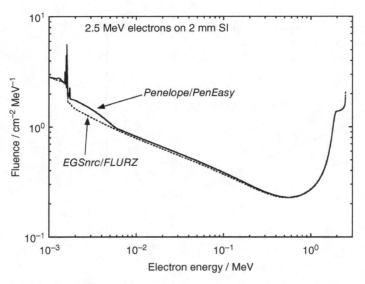

**Figure 8.9** Comparison of the spectra calculated with *EGSnrc/FLURZ* and *Penelope/PenEasy* for 2.5 MeV electrons incident on a 2 mm thick layer of silicon using $E_{cut} = \Delta = 1$ keV.

**8** Using the electron spectra calculated in the previous exercise, compute the restricted cema for $\Delta = 10$ keV, see Eq. (4.79) in the textbook. Compare the result with the absorbed dose obtained directly with a Monte Carlo simulation. (Note that some of the available MC codes can compute fluence and energy deposition during the same simulation; others require two independent simulations.)

*Answer:*
*Penelope: cavity integral = 2.1727 MeV/g; track-end term = 0.1382 MeV/g;*
$C_\Delta = 3.7025 \times 10^{-10}$ *Gy. D(MC) = 3.7047 × 10^{-10} Gy.*
*EGSnrc: cavity integral = 2.1558 MeV/g; track-end term = 0.1331 MeV/g;*
$C_\Delta = 3.6674 \times 10^{-10}$ *Gy. D(MC) = 3.7169 × 10^{-10} Gy.*

**Solution:**
The restricted cema expression is

$$C_{\text{med},\Delta} = \int_\Delta^{E_{\max}} \left[\Phi_E^{\text{tot}}\right]_{\text{med}} \left[L_\Delta(E)/\rho\right]_{\text{med}} dE + \left[\Phi_E^{\text{tot}}(\Delta)\right]_{\text{med}} \left[S_{\text{el}}(\Delta)/\rho\right]_{\text{med}} \Delta$$

where the meaning of each quantity is provided in the relevant textbook chapters. Under CPE conditions the cema is approximately equal to the absorbed dose. We refer to the first term as the 'cavity integral' and the second as the 'track-end term'. Note that for this type of calculations it is always convenient to calculate the total electron spectrum down to an energy slightly below $\Delta$, so that interpolation at this energy is possible and avoids questionable extrapolations.

The integration can be done numerically using any software for that purpose (e.g., *Mathematica* or *MatLab*) or even as a simple summation (user coded or with *Excel*). The energy at which each integral component ($\Phi_E^{\text{tot}}$ and $L_\Delta(E)/\rho$)

is evaluated is the mid-energy of each energy bin, and if a summation is made, then d$E$ must be estimated from the spectrum output bin widths.

(a) We use first the *Penelope/PenEasy* spectrum from the previous exercise. Numerical details are not given as it has 1000 energy bins, but a spectrum resembling the above should provide results within a few percent. Restricted and unrestricted stopping power data can be taken from the Data Tables or extracted from the MC system.

Our results (using *Mathematica*) are as follows:

$\Delta = 10$ keV

cavity integral $= 2.1727$ MeV/g

track-end $= 0.1382$ MeV/g [with $\Phi_E^{tot}(\Delta) = 0.8195$;

$$S_{el}(\Delta)/\rho = 16.863]^2$$

and

$$C_\Delta = 2.3109 \text{ MeV/g} = 3.7025 \times 10^{-10} \text{ Gy}$$

For the absorbed dose calculation, *Penelope/PenEasy* provides a 'tallyEnergyDeposition.dat' in eV per incident electron. We convert it into MeV/g per unit fluence dividing by $10^6$ and by the slab thickness (0.2 cm) times the silicon density (2.33 g cm$^{-3}$). Note that the units are cm$^2$ MeV/g, and per incident unit fluence (cm$^{-2}$), we get MeV/g. Hence

$$D(\text{MC}) = \frac{1.0776 \times 10^6 \text{ eV}}{10^6 \text{ eV/MeV} \times 0.2 \times 2.33} = 2.3125 \text{ MeV/g}$$

which is practically identical to the cema obtained above.

(b) We use the *EGSnrc/FLURZ* spectrum from the previous exercise. For consistency with the MC calculation, restricted and unrestricted stopping power data have been extracted using *PEGS4* and the code *examin* (see the manual).[3]

Our results (using *Mathematica*) are as follows:

$\Delta = 10$ keV

cavity integral $= 2.1558$ MeV/g

track-end $= 0.1331$ MeV/g [with $\Phi_E^{tot}(\Delta) = 0.7881$;

$$S_{el}(\Delta)/\rho = 16.892]$$

and

$$C_\Delta = 2.2890 \text{ MeV/g} = 3.6674 \times 10^{-10} \text{ Gy}$$

For the absorbed dose calculation, we have used the code *EGSnrc/DOSRZ*, which directly provides Gy per incident fluence. Hence

$$D(\text{MC}) = 3.7169 \times 10^{-10} \text{ Gy}$$

that differs from the restricted cema above by about 1.3%.

---

2 Strictly, the contribution by $S_{rad}(\Delta)/\rho$ should have been included in the track-end term, but this is equal to 0.0073, that is, negligible compared with that of $S_{el}(\Delta)/\rho$.

3 In our version of the *EGSnrc* system, we have found a small inconsistency in the silicon density between *PEGS4*, 2.4 g cm$^{-3}$, and the file ICRU 'silicon.density', 2.33 g cm$^{-3}$; we have corrected both to 2.33 g cm$^{-3}$.

Comparing the *EGSnrc/FLURZ/DOSRZ* with the *Penelope/PenEasy* calculation, the main difference is in the cavity integral (2.1558 MeV/g vs. 2.1727 MeV/g), as the track-end terms are quite similar and so are the two $D(MC)$. This small discrepancy ($\sim$1%) is most likely to be due to the spectral differences shown in Figure 8.9 of the previous exercise.

# 9

# Cavity Theory

1  A boundary region between carbon and aluminium media is traversed by a fluence of $4.10 \times 10^{11}$ electrons cm$^{-2}$ at an energy of 12.5 MeV. Ignoring δ rays and scattering, what is the absorbed dose $D_C$ in the carbon adjacent to the boundary, and what is the dose ratio $D_{Al}/D_C$?
   *Answer: 115.0 Gy, 0.947*

   **Solution:**
   The absorbed dose in carbon is given by

   $$D_C = \Phi \, (S_{el}/\rho)_{C,12.5\,\mathrm{MeV}}$$

   From the Data Tables, the mass electronic stopping power for 12.5 MeV electrons in carbon is 1.7512 MeV cm$^2$ g$^{-1}$; therefore

   $$D_C = 4.10 \times 10^{11} \, \frac{el}{cm^2} \times 1.7512 \, \frac{MeV\ cm^2}{g\ el}$$
   $$\times\ 1.602 \times 10^{-10} \, \frac{Gy}{MeV/g} = 115.02 \text{ Gy}$$

   From Eq. (9.9) in the textbook, using the mass electronic stopping power for 12.5 MeV electrons in aluminium,

   $$\frac{D_{Al}}{D_C} = \left[\frac{(S_{el}/\rho)_{Al}}{(S_{el}/\rho)_C}\right]_{12.5\ \mathrm{MeV}} = \frac{1.6580}{1.7512} = 0.9468$$

2  A small air-filled cavity ionization chamber has copper walls with thickness equal to the maximum electron range. The cavity volume is 0.100 cm$^3$, the air density is 0.0012045 g cm$^{-3}$, and a given photon exposure generates a charge (either sign) of $7.00 \times 10^{-10}$ C. (a) What is the average absorbed dose in the cavity air? (b) Apply the BG theory to estimate the absorbed dose in the adjacent copper wall, assuming a mean energy $\bar{E} = 0.43$ MeV for the cavity-crossing electrons. (c) Suppose $\bar{E}$ is 34% in error and should have the value 0.65 MeV. Redo part (b). What was the resulting percentage error in $D_{Cu}$?
   *Answer: (a) 0.1974 Gy. (b) 0.1500 Gy. (c) 0.1502 Gy, 0.2%*

*Fundamentals of Ionizing Radiation Dosimetry: Solutions to Exercises,* First Edition.
Pedro Andreo, David T. Burns, Alan E. Nahum, and Jan Seuntjens.
© 2017 Wiley-VCH Verlag GmbH & Co. KGaA. Published 2017 by Wiley-VCH Verlag GmbH & Co. KGaA.

**Solution:**

(a) In an air-filled ionization chamber, the absorbed dose to air is given by (see e.g., Eq. (10.4) or (11.5))

$$D_{air} = \frac{q}{m_{air}} \left( \frac{W}{e} \right)_{air}$$

Thus

$$D_{air} = \frac{7.00 \times 10^{-10} \text{ C} \times 33.97 \text{ J/C}}{0.100 \text{ cm}^3 \times 0.0012045 \text{ g/cm}^3 \times 1 \text{ kg/10}^3 \text{ g}} = 0.1974 \text{ Gy}$$

(b) From Eq. (9.9) and using mass electronic stopping powers from the electronic Data Tables,

$$D_{Cu} = D_{air} \left[ \frac{(S_{el}/\rho)_{Cu}}{(S_{el}/\rho)_{air}} \right]_{0.43 \text{ MeV}} = 0.1974 \times \frac{1.4185}{1.8667} = 0.1500 \text{ Gy}$$

(c)

$$D_{Cu} = D_{air} \left[ \frac{(S_{el}/\rho)_{Cu}}{(S_{el}/\rho)_{air}} \right]_{0.65 \text{ MeV}} = 0.1974 \times \frac{1.3116}{1.7235} = 0.1502 \text{ Gy}$$

that is, 0.15% error despite the large difference in mean electron energy.

3 Consider a BG cavity chamber with equilibrium-thickness copper walls. It is first filled with a mass $m$ of air and then by the same mass of hydrogen. Assuming identical photon irradiations in the two cases, what is the charge ratio $q_{air}/q_H$? Assume that $(W/e)_H = 36.5 \text{ J C}^{-1}$ and that the mean electron energy $\bar{E} = 0.80$ MeV. Note that for hydrogen $(S_{el}/\rho)_{0.8 \text{ MeV}} = 3.883 \text{ MeV cm}^2 \text{ g}^{-1}$.
*Answer:* $q_{air}/q_H = 0.466$

**Solution:**

The absorbed dose in an ionization chamber filled with two different gases, $gas_1$ and $gas_2$, can be written as

$$D_{gas_1} = D_{wall} \, s_{gas_1,wall} = \frac{q_{gas_1}}{m_{gas_1}} (W/e)_{gas_1}$$

$$D_{gas_2} = D_{wall} \, s_{gas_2,wall} = \frac{q_{gas_2}}{m_{gas_2}} (W/e)_{gas_2}$$

and from these expressions the ratio of charges for the chamber filled with air and hydrogen is given by

$$\frac{q_{air}}{q_H} = \frac{m_{air}}{m_H} \times \frac{(W/e)_H}{(W/e)_{air}} \times [s_{air,H}]_{0.8 \text{ MeV}} = 1 \times \frac{36.5}{33.97} \times \frac{1.683}{3.883} = 0.466$$

where the stopping power for air ($1.683 \text{ MeV cm}^2 \text{ g}^{-1}$) has been obtained from the electronic Data Tables.

**4**  Two air-filled cavity ionization chambers are identical except that one has an aluminium wall and the other graphite. The walls are thicker than the maximum secondary-electron range for 1 MeV photons, which are negligibly attenuated. Calculate the approximate ratio of the charge generated in the two chambers, assumed to be BG cavities.

*Answer:* $q_{Al}/q_C = 1.093$

**Solution:**
See Section 9.3.2.1 in the textbook.
Assuming that only Compton interactions occur, the mean initial electron energy from the Compton effect for photons with energy $k = 1$ MeV is obtained from $k \times \sigma_{tr}^{KN}/\sigma^{KN}$. From the photon Klein–Nishina cross sections in the Data Tables,

$$\bar{E}_0 = k \times \frac{\sigma_{tr}^{KN}}{\sigma^{KN}} = 1 \text{ MeV} \times \frac{0.9294 \times 10^{-25}}{0.2112 \times 10^{-24}} = 0.440 \text{ MeV}$$

The mean energy of the equilibrium spectrum may be crudely estimated as $\bar{E}_0/2$; thus

$$\bar{E} = \bar{E}_0/2 = 0.22 \text{ MeV}$$

For a given photon energy fluence $\Psi$, the absorbed dose to a chamber wall can be written, under CPE conditions, as

$$D_{wall} = \Psi \, (\mu_{en}/\rho)_{wall}$$

and can also be obtained as

$$D_{wall} = D_{air} \, s_{wall,air} = \frac{q}{v\rho_{air}} (W/e)_{air} \, s_{wall,air}$$

where $v$ is the chamber volume and $v\rho_{air} = m_{air}$.
From these two expressions, we obtain

$$q = \frac{\Psi \, (\mu_{en}/\rho)_{wall} \, v\rho_{air}}{(W/e)_{air} \, s_{wall,air}}$$

and for the ratio of the charges in our two chambers, identical except for their wall material, we obtain

$$\frac{q_{Al}}{q_C} = [(\mu_{en}/\rho)_{Al,C}]_{1 \text{ MeV}} \, [s_{C,Al}]_{0.22 \text{ MeV}} = \frac{0.0269}{0.0279} \times \frac{2.3657}{2.0854} = 1.093$$

**5**  (a) Compare the $\mu_e/\rho$ value obtained from Eq. (9.39) in the textbook with the value 13.4 cm$^2$ g$^{-1}$ obtained experimentally by Paliwal and Almond (1976) for the case of LiF irradiated by $^{60}$Co γ rays. (b) What constant is needed on the right-hand side of Eq. (9.39) in place of 0.01 to give exact agreement?

*Answer: 17.8 cm$^2$ g$^{-1}$, 0.031*

**Solution:**
(a) From Eq. (9.39)

$$e^{-\mu_e \, t_{max}} = 0.01$$

$$\mu_e \, t_{max} = \frac{\mu_e}{\rho} \times \rho \, t_{max} = 4.605$$

From the text following Eq. (9.39), and considering that $\bar{Z}_{LiF} = 7.5$ (see the Data Tables), we approximate $\rho \, t_{max} = 0.95 \, R_{CSDA}$, where $R_{CSDA}$ is the range (g/cm$^2$) of electrons having the mean initial energy $\bar{E}_0$ resulting from Compton interactions:

$$\bar{E}_0 = \bar{k} \times \frac{\sigma_{tr}^{KN}}{\sigma^{KN}} = 1.25 \times \frac{8.885 \times 10^{-26}}{1.888 \times 10^{-25}} = 0.588 \text{ MeV}$$

and $R_{CSDA}(0.588 \text{ MeV}) = 0.2725 \text{ g cm}^{-2}$ from the Data Tables. Therefore $\rho \, t_{max} = 0.95 \times 0.2725 = 0.2589 \text{ g cm}^{-2}$ and $\mu_e/\rho = 4.605/0.2589 = 17.8 \text{ cm}^2 \text{ g}^{-1}$:

(b)

$$\rho \, t_{max} \times \frac{\mu_e}{\rho} = 0.2589 \times 13.4 = 3.47$$

and

$$e^{-3.47} = 0.031$$

Note that Janssens *et al.* (1974) proposed 0.04.

6  A layer of water 1 mm thick between two equilibrium-thickness layers of polystyrene is irradiated by 2 MeV photons. (a) From the Burlin theory, calculate the approximate average dose in the water ($\bar{D}_w$) if the electronic kerma in the adjacent polystyrene is 10 Gy. Use Eq. (9.39) in the textbook as amended by Janssens *et al.* (1974) to obtain $\mu_e$, assume a diffuse electron field, and neglect photon attenuation. (b) What are the small-cavity (BG) and large-cavity limiting values for $\bar{D}_w$?
*Answer: (a) 10.27 Gy. (b) Small limit: 10.22 Gy, large limit:10.32 Gy*

**Solution:**
(a) Equation (9.33) in the textbook for the Burlin cavity relation can be written as

$$\frac{D_w}{(K_{el})_{poly}} = \omega_{BG}(\bar{L}_\Delta/\rho)_{poly}^w + (1 - \omega_{BG})(\bar{\mu}_{en}/\rho)_{poly}^w$$

with the data (from the electronic Data Tables)

$$(\mu_{en}/\rho)_{w,2 \text{ MeV}} = 0.0261 \text{ cm}^2 \text{ g}^{-1}$$

$$(\mu_{en}/\rho)_{poly,2 \text{ MeV}} = 0.0253 \text{ cm}^2 \text{ g}^{-1}$$

The mean energy of the electron equilibrium spectrum for the Compton effect is obtained as in previous exercises

$$\bar{E} = \bar{E}_0/2 = \left( 2 \text{ MeV} \times \frac{0.07769}{0.1464} \right) \Big/ 2 = 0.531 \text{ MeV}$$

and

$$(S_{el}/\rho)_{w,0.531 \text{ MeV}} = 2.0013 \text{ MeV cm}^2 \text{ g}^{-1}$$
$$(S_{el}/\rho)_{poly,0.531 \text{ MeV}} = 1.9583 \text{ MeV cm}^2 \text{ g}^{-1}$$

From Eq. (9.34) in the textbook

$$\omega_{BG} = \frac{1 - e^{-\mu_e \bar{\ell}}}{\mu_e \bar{\ell}}$$

where $\bar{\ell}$ is the mean chord length across the water "cavity," $\bar{\ell} = 4v/a$, $v$ being the cavity volume and $a$ its surface area. Assuming that the diameter of the water layer is much larger than its thickness $t$, $a \simeq 2 A$, and $v = A t$, where $A$ is the area of one flat surface of the water layer. Thus $\bar{\ell} \simeq 2 t = 0.2$ cm and $\rho\bar{\ell} = 0.2 \text{ g cm}^{-2}$.
Equation (9.39) with Janssens' constant is

$$e^{-\mu_e t_{max}} = 0.04$$
$$\mu_e t_{max} = \frac{\mu_e}{\rho} \times \rho t_{max} = 3.219$$

For water $\bar{Z}_w = 6.6$, and as in the previous exercise, $\rho t_{max} = 0.95 R_{CSDA}$, where $R_{CSDA}$ is the range (g/cm$^2$) in water of electrons having the mean initial energy $\bar{E}_0$ resulting from Compton interactions:

$$\bar{E}_0 = \bar{k} \times \frac{\sigma_{tr}^{KN}}{\sigma^{KN}} = 2 \text{ MeV} \times \frac{0.07769}{0.1464} = 1.061 \text{ MeV}$$

and $R_{CSDA}(1.061 \text{ MeV}) = 0.4708 \text{ g cm}^{-2}$ from the Data Tables. Thus $\rho t_{max} = 0.95 \times 0.4708 = 0.4473 \text{ g cm}^{-2}$ and

$$\frac{\mu_e}{\rho} = \frac{3.219}{0.4473} = 7.197 \frac{\text{cm}^2}{\text{g}}$$

Hence

$$\omega_{BG} = \frac{1 - e^{-7.197 \times 0.2}}{7.197 \times 0.2} = 0.530$$

and

$$D_w = (K_{el})_{poly}[\omega_{BG}(\bar{L}_\Delta/\rho)_{poly}^w + (1 - \omega_{BG})(\bar{\mu}_{en}/\rho)_{poly}^w]$$

$$= 10 \text{ Gy} \left[0.530 \times \frac{2.0013}{1.9583} + (1 - 0.530) \times \frac{0.0261}{0.0253}\right] = 10.267 \text{ Gy}$$

N.B. The differences between the BG and the SA stopping-power ratios are assumed to be negligible.

(b) For the small-cavity limit ($\omega_{BG} = 1$),

$$D_w = 10 \text{ Gy} \times \frac{2.0013}{1.9583} = 10.22 \text{ Gy}$$

For the large-cavity limit ($\omega_{BG} = 0$),

$$\bar{D}_w = 10 \text{ Gy} \times \frac{0.0261}{0.0253} = 10.32 \text{ Gy}$$

**7** A dilute aqueous chemical dosimeter (assumed to be similar to water) is enclosed in an equilibrium-thickness capsule of PMMA and exposed to $^{60}$Co γ rays in a water phantom. Calculate the approximate ratio of $\bar{D}_{det}$ in this dosimeter to the dose under CPE conditions in water, $D_w$, assuming the weighting factor in Burlin's theory to be firstly $\omega_{BG} = 1$ and secondly $\omega_{BG} = 0$. Ignore any other parameter related to the detector cavity size, that is, take $L_\Delta/\rho \approx S_{el}/\rho$.

*Answer: $\bar{D}_{det}/D_w = 0.9966$ for $\omega_{BG} = 1$, and $\bar{D}_{det}/D_w = 1.000$ for $\omega_{BG} = 0$*

**Solution:**

We assume that photons interact only by the Compton effect.

For $^{60}$Co γ rays ($\bar{k} = 1.25$ MeV), the mass energy-absorption coefficients in water and in PMMA are (see the electronic Data Tables)

$$(\mu_{en}/\rho)_w = 0.0296 \text{ cm}^2\,\text{g}^{-1}$$

$$(\mu_{en}/\rho)_{PMMA} = 0.0288 \text{ cm}^2\,\text{g}^{-1}$$

The mean initial electron energy from the Compton effect for $\bar{k} = 1.25$ MeV is obtained from $\bar{k} \times \sigma_{tr}^{KN}/\sigma^{KN}$. From the photon Klein–Nishina cross sections in the Data Tables,

$$\bar{E}_0 = \bar{k} \times \frac{\sigma_{tr}^{KN}}{\sigma^{KN}} = 1.25 \times \frac{8.885 \times 10^{-26}}{1.888 \times 10^{-25}} = 0.588 \text{ MeV}$$

The mean energy in the equilibrium 'slowing-down spectrum' of the secondary electrons created by the Compton electrons is $\bar{E} \approx 0.5 \times \bar{E}_0 = 0.294$ MeV, and for this energy (see the electronic Data Tables)

$$(S_{el}/\rho)_w = 2.3623 \text{ MeV cm}^2\,\text{g}^{-1}$$

$$(S_{el}/\rho)_{PMMA} = 2.3100 \text{ MeV cm}^2\,\text{g}^{-1}$$

The ratio $\bar{D}_{det}/D_w$ can be written as

$$\frac{\bar{D}_{det}}{D_w} = \frac{\bar{D}_{det}}{D_{PMMA}} \times \frac{D_{PMMA}}{D_w}$$

where, from the Burlin cavity theory, and noting that unrestricted stopping powers are used,

$$\frac{\bar{D}_{det}}{D_{PMMA}} = \omega_{BG} \frac{2.3623}{2.3100} + (1 - \omega_{BG}) \frac{0.0296}{0.0288}$$

$$= 1.0226\,\omega_{BG} + 1.0261\,(1 - \omega_{BG})$$

and, under CPE conditions,

$$\frac{D_{PMMA}}{D_w} = (\mu_{en}/\rho)_{PMMA,w}$$

Hence

$$\frac{\bar{D}_{det}}{D_w} = \frac{\bar{D}_{det}}{D_{PMMA}} \times (\mu_{en}/\rho)_{PMMA,w}$$

$$= \left[\omega_{BG} \frac{2.3623}{2.3100} + (1 - \omega_{BG}) \frac{0.0296}{0.0288}\right] \times \frac{0.0288}{0.0296}$$

$$= \omega_{BG} \frac{2.3623}{2.3100} \times \frac{0.0288}{0.0296} + (1 - \omega_{BG})$$

$$= 0.9966 \, \omega_{BG} + (1 - \omega_{BG}) = 1 - 0.0034 \, \omega_{BG}$$

Thus $\bar{D}_{det}/D_w$ is equal to 0.9966 when $\omega_{BG} = 1$, and 1.000 for $\omega_{BG} = 0$. PMMA walls can be seen to provide a very close (0.34 %) match to the aqueous solution in the dosimeter's sensitive volume for $^{60}$Co $\gamma$ rays.

**8** The cut-off ($\Delta$) for Spencer–Attix stopping-power ratios for ionization chambers in radiotherapy dosimetry is usually taken to be 10 keV. Discuss the meaning of this cut-off. Taking as examples a plane-parallel chamber (radius = 10 mm; height = 1 mm) and a cylindrical Farmer-type chamber (radius = 3.5 mm; length = 24 mm), discuss whether 10 keV is a reasonable choice in both cases. The mean chord length of a cylinder is approximated by $4 \times$ volume/area in both cases. Electron stopping power data and related quantities for air are given in the electronic Data Tables.

**Solution**
In the Spencer–Attix formulation, the total electron fluence entering the cavity is divided into two components. Electrons with energies below the cut-off energy $\Delta$ are assumed to deposit all of their energy in the cavity; this is the 'track-end' term introduced by Nahum. Electrons with energies above $\Delta$ lose energy by liberating electrons with energies above and below $\Delta$; those with energy below $\Delta$ are also assumed to deposit all of their energy in the cavity, while those above $\Delta$ are assumed to escape the cavity without depositing any energy. The fraction of the total energy loss given to electrons below $\Delta$ is $L(E, \Delta)/S_{el}(E)$, which effectively defines the restricted stopping power $L(E, \Delta)$. Thus, energy deposition in the cavity is assumed to be entirely mediated by electrons with energies up to $\Delta$, from which it is evident that $\Delta$ is related to the size of cavity.

To estimate if 10 keV is a reasonable figure, we need to determine the mean chord length in each case, and from the table below, we obtain the electron energy whose $R_{CSDA}$ corresponds to such a distance.

| | Plane-parallel chamber | Farmer chamber |
|---|---|---|
| Radius (cm) | 1.0 | 0.35 |
| Height or length (cm) | 0.1 | 2.40 |
| Volume ($\pi \, r^2 h$) (cm$^3$) | 0.314 | 0.924 |
| Area base ($\pi \, r^2$) (cm$^2$) | 3.142 | 0.385 |
| Area side ($2 \, \pi \, r \, h$) (cm$^2$) | 0.628 | 5.278 |
| Area total (2 base + side) (cm$^2$) | 6.912 | 6.048 |
| Mean chord length, $\bar{\ell} = 4v/a$ (cm) | 0.182 | 0.611 |
| Mean chord length (g cm$^{-2}$) | $2.191 \times 10^{-4}$ | $7.361 \times 10^{-4}$ |
| Electron energy for $R_{CSDA}$(air)= $\bar{\ell}$ | 9 keV | 17 keV |

This means that for the plane-parallel chamber, the cut-off value appears to be reasonable, but it is somewhat low for the Farmer chamber.

**9** With reference to Fano's theorem, explain why ionization chambers were originally designed with graphite walls. Explain also the rationale for their current construction with plastic walls for high-energy beams.

**Solution**
Fano's theorem states, *"In an infinite medium of given atomic composition exposed to a uniform fluence of primary radiation (such as x rays or neutrons), the fluence of secondary radiation (electrons) is also uniform and independent of the density of the medium, as well as of density variations from point to point."*
The Fano Theorem applies to a single medium. However, an ionization chamber in a phantom presents a three-material problem: the phantom material, the chamber wall, and the cavity gas, normally air. Reducing this to a single-material problem depends on the photon energy range:

- At low photon energies, photon interactions in the cavity gas compete with those in the chamber wall. If these two materials are similar then the electron spectrum remains undisturbed. Graphite was chosen as a good approximation to air, and subsequently air-equivalent plastics were developed.
- At high photon energies, photon interactions in the cavity are negligible, but photon interactions in the phantom compete with those in the chamber wall. In this case, the similarity of the wall and phantom material is important (unless the wall is very thin). For a water phantom, graphite again can fulfill this role, as well as water-equivalent plastic. In air, graphite and air-equivalent plastic are suitable.

# 10

# Overview of Radiation Detectors and Measurements

**1**  Name four purposes that the wall of a dosimeter may serve.
*Answer: (1) Defines the interior volume; (2) provides CPE; (3) stops charged particles entering from outside and (4) modifies the energy dependence.*

**2**  Plot the photon energy dependence, relative to $^{60}$Co γ rays, of a (large) detector response per unit of dose to air for carbon, silicon, PMMA, polystyrene, and LiF TLD.
*Answer: The reading of a 'large photon' detector is proportional to $\mu_{en}/\rho$. Use data from the Data Tables to plot the ratio $(\mu_{en}/\rho)_{med,air}$ relative to that at 1.25 MeV.*

**Solution:**
In photons, the detector response per unit of dose to air can be estimated from the ratio of mass energy-absorption coefficients $(\mu_{en}/\rho)_{med,air}$ so that, relative to 1.25 MeV, the detector response is given by

$$R \approx \frac{\left[ \dfrac{(\mu_{en}/\rho)_{med}}{(\mu_{en}/\rho)_{air}} \right]_k}{\left[ \dfrac{(\mu_{en}/\rho)_{med}}{(\mu_{en}/\rho)_{air}} \right]_{1.25}}$$

and is plotted in Figure 10.1 below.

**3**  A LiF TL dosimeter is enclosed in equilibrium-thickness capsules of polystyrene and PMMA for γ-ray energies of 0.20 MeV and 1.25 MeV. Calculate the ratio of the average dose in the LiF in the dosimeter to the LiF dose under CPE conditions for each capsule and energy, assuming $\omega_{BG} = 1$ and 0 in the Burlin theory. Which capsule gives the closer match to the LiF in the dosimeter?

*Fundamentals of Ionizing Radiation Dosimetry: Solutions to Exercises,* First Edition.
Pedro Andreo, David T. Burns, Alan E. Nahum, and Jan Seuntjens.

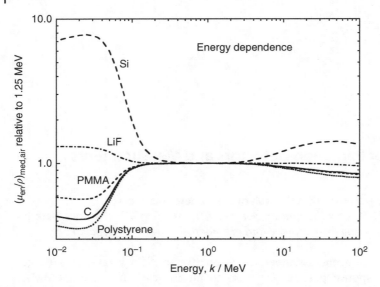

**Figure 10.1** Ratio of mass energy-absorption coefficient materials to air, relative to the ratio at 1.25 MeV.

*Answer: For $\omega_{BG} = 1$:*

| Material | Energy (MeV) | $\bar{D}_{det}/D_{LiF}^{CPE}$ |
|---|---|---|
| Polystyrene | 0.20 | 0.937 |
| | 1.25 | 0.961 |
| PMMA | 0.20 | 0.950 |
| | 1.25 | 0.972 |

*For $\omega_{BG} = 0$ in all cases $\bar{D}_{det}/D_{LiF} = 1.000$.*
*For PMMA the dose ratio is slightly closer to unity, that is, it is a better match.*

**Solution:**
Burlin's expression (Eq. (9.33) in the textbook) can be written as

$$\frac{\bar{D}_{det}}{D_{LiF}^{CPE}} = \frac{\bar{D}_{det}}{D_{pol}}\frac{D_{pol}}{D_{LiF}^{CPE}} = \frac{\bar{D}_{det}}{D_{pol}} \times (\mu_{en}/\rho)_{LiF}^{pol}$$

$$= \left[\omega_{BG}\,(S_{el}/\rho)_{pol}^{LiF} + (1 - \omega_{BG})(\mu_{en}/\rho)_{pol}^{LiF}\right] \times (\mu_{en}/\rho)_{LiF}^{pol}$$

and the same equation can be written for PMMA instead of polystyrene. Assuming that only the Compton effect is important in low-$Z$ media at 0.2 MeV or 1.25 MeV, the average equilibrium spectrum secondary electron energies are approximately given by:
For $k = 0.2$ MeV:

$$\bar{E} \approx \bar{E}_0/2 = \frac{0.2}{2}\frac{0.8794}{4.065} = 2.2 \times 10^{-2} \text{ MeV} = 0.02 \text{ MeV}$$

For $k = 1.25$ MeV:

$$\bar{E} \approx \frac{1.25}{2} \frac{0.8885}{1.888} \approx 0.3 \text{ MeV}$$

From the Data Tables:

| Material | $(S_{el}/\rho)_{0.02\text{ MeV}}$ | $(S_{el}/\rho)_{0.3\text{ MeV}}$ | $(\mu_{en}/\rho)_{0.2\text{ MeV}}$ | $(\mu_{en}/\rho)_{1.25\text{ MeV}}$ |
|---|---|---|---|---|
| LiF | 10.55 | 1.907 | 0.0248 | 0.0247 |
| Polystyrene | 12.96 | 2.305 | 0.0286 | 0.0287 |
| PMMA | 12.83 | 2.292 | 0.0287 | 0.0288 |

For $k = 0.2$ MeV and polystyrene capsule:

$$\frac{\bar{D}_{det}}{D_{LiF}^{CPE}} = \left[ \omega_{BG} \frac{10.55}{12.96} + (1 - \omega_{BG}) \frac{0.0248}{0.0286} \right] \times \frac{0.0286}{0.0248} = 0.937 \ \omega_{BG} + (1 - \omega_{BG})$$

that is, 0.937 for $\omega_{BG} = 1$ or 1.000 for $\omega_{BG} = 0$.
For $k = 1.25$ MeV and polystyrene capsule:

$$\frac{\bar{D}_{det}}{D_{LiF}^{CPE}} = \left[ \omega_{BG} \frac{1.907}{2.305} + (1 - \omega_{BG}) \frac{0.0247}{0.0287} \right] \times \frac{0.0287}{0.0247} = 0.961 \ \omega_{BG} + (1 - \omega_{BG})$$

that is, 0.961 for $\omega_{BG} = 1$ or 1.000 for $\omega_{BG} = 0$.
For $k = 0.2$ MeV and PMMA capsule:

$$\frac{\bar{D}_{det}}{D_{LiF}^{CPE}} = \left[ \omega_{BG} \frac{10.55}{12.83} + (1 - \omega_{BG}) \frac{0.0248}{0.0287} \right] \times \frac{0.0287}{0.0248} = 0.950 \ \omega_{BG} + (1 - \omega_{BG})$$

that is, 0.950 for $\omega_{BG} = 1$ or 1.000 for $\omega_{BG} = 0$.
For $k = 1.25$ MeV and PMMA capsule:

$$\frac{\bar{D}_{det}}{D_{LiF}^{CPE}} = \left[ \omega_{BG} \frac{1.907}{2.292} + (1 - \omega_{BG}) \frac{0.0247}{0.0288} \right] \times \frac{0.0288}{0.0247} = 0.972 \ \omega_{BG} + (1 - \omega_{BG})$$

that is, 0.972 for $\omega_{BG} = 1$ or 1.000 for $\omega_{BG} = 0$.
Therefore PMMA gives $\bar{D}_{det}/D_{LiF}^{CPE}$ slightly closer to unity, that is, it is a better match.

**4** A LiF dosimeter is enclosed in an equilibrium-thickness wall of aluminium to determine the dose to silicon under PCPE conditions in a field of 6 MeV γ rays. Neglecting x-ray attenuation, if the average dose in the LiF is 30 Gy, what is the silicon dose? Assume that the LiF is (a) small (the Burlin $\omega_{BG} = 1$) or (b) large ($\omega_{BG} = 0$).
*Answer: (a) 31.29 Gy, (b) 35.81 Gy*

**Solution:**
(a) Use the same approach as in the previous exercise for the Burlin expression and consider that for a small cavity ($\omega_{BG} = 1$) we can write

$$\frac{\bar{D}_{det}}{D_{Si}} = \frac{\bar{D}_{det}}{D_{Al}}\frac{D_{Al}}{D_{Si}} = (S_{el}/\rho)_{Al}^{LiF}(\mu_{en}/\rho)_{Si}^{Al}$$

where "det" represents a LiF-filled cavity.

The average energy of the equilibrium Compton-electron spectrum is

$$\bar{E} \approx \bar{E}_0/2 = \frac{1}{2}\left(6 \times \frac{0.4716}{0.7323}\right) = \frac{3.864}{2} = 1.93 \text{ MeV}$$

Note that in the Data Tables, one can see that the pair production cross section for Al is 19% of the total, Compton being nearly all of the remainder (81%). The mean starting energy of pairs is

$$\bar{E}_0 = \frac{6 - 1.022}{2} = 2.49 \text{ MeV}$$

and their mean equilibrium energy is $\bar{E} \approx 1.25$ MeV. There are two electrons ($e^+$ and $e^-$) for each pair interaction but only one recoil electron for each Compton event. Hence the weighted mean equilibrium electron energy in Al exposed to 6 MeV γ rays is

$$\bar{E} = \frac{0.81 \times 1.93 + 2 \times 0.19 \times 1.25}{0.81 + 2 \times 0.19} = 1.7 \text{ MeV}$$

This will be nearly correct for Si also.
From the Data Tables:

| Material | $(S_{el}/\rho)_{1.7\,MeV}$ | $(\mu_{en}/\rho)_{6\,MeV}$ |
|----------|---------------------------|----------------------------|
| LiF | 1.471 | 0.0152 |
| Al | 1.465 | 0.0174 |
| Si | — | 0.0182 |

Hence

$$\frac{\bar{D}_{det}}{D_{Si}} = \frac{1.471}{1.465} \times \frac{0.0174}{0.0182} = 0.959$$

and therefore

$$D_{Si} = \frac{\bar{D}_{det}}{0.959} = \frac{30\,\text{Gy}}{0.959} = 31.29\,\text{Gy}$$

(b) Proceeding as in (a), for a large cavity ($\omega_{BG} = 0$),

$$\frac{\bar{D}_{det}}{D_{Si}} = (\mu_{en}/\rho)_{Si}^{LiF} = \frac{0.0152}{0.0182} = 0.838$$

and therefore

$$D_{Si} = \frac{\bar{D}_{det}}{0.838} = \frac{30\,\text{Gy}}{0.838} = 35.81\,\text{Gy}$$

**5** Suppose the LiF in the dosimeter in the previous exercise is 2 mm thick and the aluminium wall is just equal to the electron range. (a) Estimate the x-ray attenuation to reach the dosimeter center using the straight-ahead approximation. (b) If it were immersed in water, what would be the corresponding attenuation in the water it displaced?
*Answer: (a) 6.3%, (b) 2.5%.*

**Solution:**
The maximum energy of Compton recoil electrons is

$$E_{max} = \frac{2\,k^2}{2\,k + 0.511} = 5.75 \text{ MeV}$$

and from the Data Tables $R_{CSDA}(\text{Al}, 5.75 \text{ MeV}) = 3.53 \text{ g/cm}^2$.
(Note that the exercise says "range," not maximum penetration depth, $t_{max}$, which is about 10% less for aluminium.)
From the Data Tables, $\rho_{LiF} = 2.635 \text{ g/cm}^3$, and $0.1 \text{ cm} \times 2.635 = 0.2635 \text{ g/cm}^2$ is the mass depth of the center of the LiF chip from one of its flat sides.
At 6 MeV, from the electronic Data Tables

$$(\mu_{en}/\rho)_{Al} = 0.0174 \text{ cm}^2/\text{g}$$
$$(\mu_{en}/\rho)_w = 0.0180 \text{ cm}^2/\text{g}$$
$$(\mu_{en}/\rho)_{LiF} = 0.0152 \text{ cm}^2/\text{g}$$

(a) The energy fluence attenuation is calculated from the straight-ahead approximation (see Section 5.4 in the textbook) according to

$$\frac{\Psi_{ctre}}{\Psi_0} = \exp\left[-(\mu_{en}/\rho)_{LiF}\,(\rho\,t)_{LiF} - (\mu_{en}/\rho)_{Al}\,(\rho\,t)_{Al}\right]$$
$$= \exp\left[-0.0152 \times 0.2635 - 0.0174 \times 3.53\right] = 0.937$$

or 6.3% attenuation.

(b) From the Data Tables, $\rho_{Al} = 2.699 \text{ g/cm}^3$; therefore

$$\frac{R_{CSDA}}{\rho} = \frac{3.53 \text{ g/cm}^2}{2.699 \text{ g/cm}^3} = 1.31 \text{ cm}$$

Total water thickness to center: $1.31 + 0.1 = 1.41 \text{ cm} = 1.41 \text{ g/cm}^2$ for $\rho_w = 1$. Thus

$$\frac{\Psi_{ctre}}{\Psi_0} = \exp\left[-0.0180 \times 1.41\right] = 0.975$$

that is, 2.5% attenuation in water.

**6** A LiF TLD chip gives a reading $M$ after receiving an absorbed dose of 4.5 Gy from $^{60}$Co $\gamma$ rays. Later, the same reading is obtained after the chip is irradiated by a beam of 500 keV electrons on its $3.2 \times 3.2 \text{ mm}^2$ face. The chip thickness is 0.9 mm. What is the electron fluence, neglecting backscattering?
*Answer: $1.33 \times 10^{10}$ electrons $cm^{-2}$*

**Solution:**

From the Data Tables, $\rho_{LiF} = 2.635$ g/cm$^3$; therefore the chip mass thickness is $2.635 \times 0.09 = 0.237$ g/cm$^2$. As the range of 500 keV electrons in LiF is 0.218 g/cm$^2$, it can be assumed that all the electrons stop in the chip.

The mass of the chip is $\approx 2.635 \times 0.09 \times 0.32 \times 0.32 = 0.0243$ g,

The absorbed dose is 4.5 Gy $\approx 4.5 \times 10^4$ erg/g,

Hence the observed reading corresponds to an imparted energy of

$$4.5 \times 10^4 \times 0.0243 = 1094 \text{ erg}$$

that is,

$$1094 \text{ erg} \times \frac{1 \text{ MeV}}{1.602 \times 10^{-6} \text{ erg}} = 6.83 \times 10^8 \text{ MeV}$$

Therefore the incident energy fluence is

$$\Psi = \frac{6.83 \times 10^8 \text{ MeV}}{(0.32 \text{ cm})^2} = 6.67 \times 10^9 \frac{\text{MeV}}{\text{cm}^2}$$

and the electron fluence is

$$\Phi = \frac{6.67 \times 10^9 \frac{\text{MeV}}{\text{cm}^2}}{0.5 \frac{\text{MeV}}{\text{el}}} = 1.33 \times 10^{10} \frac{\text{el}}{\text{cm}^2}$$

# 11

# Primary Radiation Standards

**1** A standard free-air ionization chamber has a diaphragm aperture 1.30 cm in diameter and a collector plate 12.0 cm long, separated from the guard plates by 0.5 mm gaps. The distance from the diaphragm to the edge of the guard plate is 30 cm. The dry air in the chamber is at 23.1 °C, 103.3 kPa. The effective energy of a photon beam is estimated to be 80 keV and for this energy $k_{sc} = 1.004$, $k_e = 1.001$, and $k_d = 1.002$. Calculate the air kerma at the diaphragm for a charge of $6.172 \times 10^{-7}$ C (corrected for ion recombination). Note that, because of the way in which the gaps disturb the electric field lines, the collecting length is taken to include half of the front and rear gaps (thus 12.1 cm).
*Answer: 1.090 Gy.*

**Solution:**
The basic expression for the determination of air kerma using a free-air chamber, including correction factors, can be written as (see Eq. (10.4) or (11.5) of the textbook)

$$K_{air} = \frac{q}{m_{air}} \frac{W_{air}}{e} \frac{1}{1 - g_{air}} \prod_i k_i$$

where
– $q$ is the measured charge
– $m_{air}$ is the mass of the theoretical measuring air volume ($\rho_{air} \, \pi \, r_{ap}^2 \, L$, with $\rho_{air} = 1.2045 \times 10^{-3}$ g cm$^{-3}$)
– $W_{air}/e$ is the mean energy required to produce an ion pair in air, per electron charge (33.97 J C$^{-1}$)
– $1 - g_{air}$ accounts for the energy lost in radiative processes
– $k_i$ are correction factors, namely, for attenuation ($k_a$), scatter ($k_{sc}$), electron loss ($k_e$), and diaphragm transmission and scatter ($k_d$). The former needs to be determined.

The attenuation correction factor can be estimated from the air attenuation in the distance between the diaphragm and the collector center, that is,

$$k_a = \exp(+\mu/\rho \times \rho \, l)$$

where $l = 30$ cm + 0.05 cm + 12/2 cm = 36.05 cm. Note the '+' sign, as we have to increase the measured current because attenuation has taken place.

*Fundamentals of Ionizing Radiation Dosimetry: Solutions to Exercises,* First Edition.
Pedro Andreo, David T. Burns, Alan E. Nahum, and Jan Seuntjens.

From the Data Tables, for 80 keV photons we get $(\mu/\rho)_{air} = 0.16664 \text{ cm}^2 \text{ g}^{-1}$, $1 - g_{air} = 0.99985$. This yields $k_a = 1.00726$, and therefore

$$\prod_i k_i = k_a \, k_{sc} \, k_e \, k_d = 1.01433$$

The charge needs to be corrected for temperature and pressure (see Eqs (12.6) and (12.7)), where $T_{ref} = 20\,°C$ and $P_{ref} = 101.325 \text{ kPa}$, that is,

$$k_T = \frac{273.15 + T\,(°C)}{273.15 + T_{ref}\,(°C)} = \frac{273.15 + 23.1}{273.15 + 20} = 1.0106$$

and

$$k_P = \frac{P_{ref}}{P} = \frac{101.325}{103.3} = 0.9809$$

Thus

$$q = 6.172 \times 10^{-7} \times k_T \times k_P = 6.172 \times 10^{-7} \times 1.0106 \times 0.9809$$
$$= 6.118 \times 10^{-7} \text{ C}$$

And, finally, the air mass is

$$m_{air} = 1.2045 \times 10^{-3} \frac{\text{g}}{\text{cm}^3} \times \pi \times \left(\frac{1.30}{2}\right)^2 \text{cm}^2 \times 12.1 \text{ cm}$$
$$= 19.35 \times 10^{-3} \text{ g} = 19.35 \times 10^{-6} \text{ kg}$$

The air kerma therefore results as follows:

$$K_{air} = \frac{6.118 \times 10^{-7} \text{ C}}{19.35 \times 10^{-6} \text{ kg}} \times 33.97 \frac{\text{J}}{\text{C}} \times \frac{1}{0.99985} \times 1.01433 = 1.090 \text{ Gy}$$

**2** An air-filled spherical ionization chamber has a volume of 10 cm$^3$ and a polystyrene wall thick enough to yield CPE but thin enough to disregard photon attenuation in a $^{60}$Co γ-ray beam (taken to be 1.25 MeV). The chamber will be exposed to an air-kerma rate of 50 mGy h$^{-1}$ in $^{60}$Co. Calculate the ionization current that is measured with the chamber at a temperature of 22 °C and pressure of 100 kPa. Demonstrate that the chamber behaves as a Bragg–Gray cavity; for the estimation of the mean electron energy assume that only Compton interactions are of importance and that Compton electrons receive on average half of the energy transferred by the incident photons.
*Answer: 4.673 pA.*

**Solution:**
We estimate first the mean energy of the electron equilibrium spectrum, approximated by 1/2 of the mean energy transferred in a Compton interaction. From the Data Tables for the Klein–Nishina cross sections (Table A.8), we get

$$_e\sigma_{tr} = 8.885 \times 10^{-26} \quad \text{and} \quad _e\sigma = 1.888 \times 10^{-25}$$

both in cm$^2$ g$^{-1}$. Their ratio is $_e\sigma_{tr}/_e\sigma = 0.4706$.

The mean energy transferred is therefore $1.25 \times 0.4706 = 0.588$ MeV, and the mean energy of the electron spectrum becomes $\bar{E} = 0.588/2 = 0.294$ MeV.

With regard to the sphere volume, 10 cm$^3$, its radius is 1.34 cm and its area $(4\pi r^2)$ 22.45 cm$^2$. This results in a mean chord length $(\bar{\ell} = 4$ volume/area) of 1.78 cm, which for air $(\rho_{air} = 1.2045 \times 10^{-3}$ g cm$^{-3})$ becomes $\bar{\ell} = 2.15 \times 10^{-3}$ g cm$^{-2}$. The air mass inside the chamber cavity is $m_{air} = \rho_{air} \times$ volume.

The $R_{CSDA}$ of 0.294 MeV electrons (from the Data Tables) is 0.0924 g cm$^{-2}$, that is, $R_{CSDA} \gg \bar{\ell}$ ($\sim 40$ times). Thus we can assume that the chamber behaves as a Bragg–Gray cavity, that is,

$$\dot{D}_{air} = \dot{D}_{wall} \, s_{air,wall}$$

On the other hand, as CPE exists for the photon field ($k = 1.25$ MeV),

$$\dot{D}_{wall} = \dot{K}_{air} \, (1 - \bar{g}_{air}) \, [\mu_{en}(k)/\rho]_{wall,air}$$

We have therefore

$$\dot{D}_{air} = \dot{K}_{air} \, (1 - \bar{g}_{air}) \, [\mu_{en}(k)/\rho]_{wall,air} \, s_{air,wall}$$

which can be compared with Eq. (11.11) in the textbook for the air-kerma measurement with a thick-walled ionization chamber.

The relevant data for the various terms are

(a) for photons at $k = 1.25$ MeV, $1 - \bar{g}_{air} = 0.9962$, $[\mu_{en}(k)/\rho]_{polyst} = 0.0287$ cm$^2$ g$^{-1}$ and $[\mu_{en}(k)/\rho]_{air} = 0.0267$ cm$^2$ g$^{-1}$

(b) for electrons at $\bar{E} = 0.294$ MeV, $[S_{el}(\bar{E})/\rho]_{polyst} = 2.3229$ MeV cm$^2$ g$^{-1}$ and $[S_{el}(\bar{E})/\rho]_{air} = 2.0998$ MeV cm$^2$ g$^{-1}$

resulting in an absorbed dose rate to air

$$\dot{D}_{air} = 50 \text{ mGy h}^{-1} \times 0.9962 \times \frac{0.0287}{0.0267} \times \frac{2.0998}{2.3229}$$
$$= 48.40 \text{ mGy h}^{-1} = 1.344 \times 10^{-5} \text{ Gy s}^{-1}$$

Using the basic absorbed dose expression for an ionization chamber,

$$\dot{D}_{air} = \frac{i}{m_{air}} \frac{W_{air}}{e}$$

results in

$$i = \frac{\dot{D}_{air} \, m_{air}}{W_{air}/e} = \frac{1.344 \times 10^{-5} \text{ Gy/s} \times 1.2045 \times 10^{-5} \text{ kg}}{33.97 \text{ J/C}} = 4.767 \times 10^{-12} \text{ A}$$

This needs to be corrected for temperature and pressure, that is,

$$k_T = \frac{273.15 + T(^\circ C)}{273.15 + T_{ref}(^\circ C)} = \frac{273.15 + 22}{273.15 + 20} = 1.0068$$

$$k_P = \frac{P_{ref}}{P} = \frac{101.325}{100} = 1.0133$$

resulting in (note we are going from reference to actual conditions using factors designed to go from actual to reference conditions)

$$i = \frac{4.767 \times 10^{-12} \text{ A}}{k_T \times k_P} = \frac{4.767 \times 10^{-12} \text{ A}}{1.0068 \times 1.0133} = 4.673 \text{ pA}$$

**3** An aluminium thick-walled spherical ionization chamber has a volume of $1 \text{ cm}^3$. The wall is thick enough to yield CPE but thin enough to disregard photon attenuation. Measurements are made at a facility at sea level at an air-kerma rate of $1.0 \text{ Gy h}^{-1}$ in a 412 keV $\gamma$-ray beam (from a $^{198}\text{Au}$ source), at a temperature of $22 \,°\text{C}$ and air pressure $103.5 \text{ kPa}$. The chamber is subsequently used in a facility with the same photon energy, situated at an altitude of around $3600 \text{ m}$, at a temperature of $20 \,°\text{C}$ and pressure of $60.0 \text{ kPa}$. Determine the difference in current measured in the two cases. As before, demonstrate that the chamber behaves as a Bragg–Gray cavity. *Answer: 4.63 pA.*

**Solution:**
As in the previous exercise, we estimate first the mean energy of the electron equilibrium spectrum, approximated by 1/2 of the mean energy transferred in a Compton interaction. From the Data Tables for the Klein–Nishina cross sections (Table A.8) for 412 keV,

$$_e\sigma_{tr} = 9.814 \times 10^{-26} \quad \text{and} \quad _e\sigma = 3.129 \times 10^{-25}$$

both in $\text{cm}^2 \text{ g}^{-1}$. Their ratio is $_e\sigma_{tr}/_e\sigma = 0.3136$.

The mean energy transferred is therefore $0.412 \times 0.3136 = 0.129$ MeV, and the mean energy of the electron spectrum becomes $\bar{E} = 0.129/2 = 0.065$ MeV.

For the chamber, following the procedure in the previous exercise, we get a mean chord length $\bar{\ell} = 0.001 \text{ g cm}^{-2}$, which is considerably smaller than the $R_{CSDA}(65 \text{ keV}) = 0.0077 \text{ g cm}^{-2}$. Thus Bragg–Gray criterion can be applied. From the previous exercise, the ionization chamber current is given by

$$i = \frac{\dot{D}_{air} \, m_{air}}{W_{air}/e}$$

with

$$\dot{D}_{air} = \dot{K}_{air} \, (1 - \bar{g}_{air}) \, (\mu_{en}(k)/\rho)_{wall,air} \, s_{air,wall}$$

that is, the current is proportional to the mass of air and therefore

$$\Delta i = \frac{\dot{D}_{air} \, \Delta m_{air}}{W_{air}/e}$$

To determine the change in air mass ($m_{air} = \rho_{air} \times$ volume), we should recall that the air density dependence is given by

$$\rho_{air}(T, P) = \frac{\rho_{air}(T_{ref}, P_{ref})}{k_T \, k_P}$$

where $k_T$ and $k_p$ have their usual definitions (note that the $P$ and $T$ corrections for density become the reverse of those used for charge or current corrections), that is,

$$m_{air} = \frac{\rho_{air}(T_{ref}, P_{ref})}{\frac{273.15+T(^\circ C)}{273.15+T_{ref}(^\circ C)} \times \frac{P_{ref}}{P}} \times volume$$

Thus

$$m_{air}(1) = \frac{1.2045 \times 10^{-3} \text{ g/cm}^3}{\frac{273.15+22}{273.15+20} \times \frac{101.325}{103.5}} \times 1 \text{ cm}^3 = 1.222 \times 10^{-3} \text{ g}$$

$$= 1.222 \times 10^{-6} \text{ kg}$$

$$m_{air}(2) = \frac{1.2045 \times 10^{-3} \text{ g/cm}^3}{\frac{273.15+20}{273.15+20} \times \frac{101.325}{60}} \times 1 \text{ cm}^3 = 7.133 \times 10^{-4} \text{ g}$$

$$= 7.133 \times 10^{-7} \text{ kg}$$

and therefore $\Delta m_{air} = 5.088 \times 10^{-7}$ kg.
For the $\dot{K}_{air} = 1$ Gy h$^{-1}$ and taking into account the data,
(a) for photons at $k = 0.412$ MeV, $1 - \bar{g}_{air} = 0.9991$, $[\mu_{en}(k)/\rho]_{Al} = 0.0286$ cm$^2$ g$^{-1}$ and $[\mu_{en}(k)/\rho]_{air} = 0.0295$ cm$^2$ g$^{-1}$
(b) for electrons at $\bar{E} = 0.065$ MeV, $[S_{el}(\bar{E})/\rho]_{Al} = 4.2216$ MeV cm$^2$ g$^{-1}$ and $[S_{el}(\bar{E})/\rho]_{air} = 4.8550$ MeV cm$^2$ g$^{-1}$
the dose rate becomes

$$\dot{D}_{air} = 1 \text{ Gy/h} \times 0.9991 \times \frac{0.0286}{0.0295} \times \frac{4.8550}{4.2216} = 1.114 \text{ Gy h}^{-1}$$

This results in the following:

$$\Delta i = \frac{\dot{D}_{air} \Delta m_{air}}{W_{air}/e} = \frac{1.114 \text{ Gy/h} \times 5.088 \times 10^{-7} \text{ kg}}{33.97 \text{ J/C}}$$

$$= 4.635 \times 10^{-12} \text{ A} = 4.635 \text{ pA}$$

**4** A standard ionization chamber was placed at a depth of 30 mm in a water phantom and irradiated with 20 MeV electrons from a linear accelerator. The accelerator monitor had been calibrated to deliver a dose to water of 1.00 Gy per 100 monitor units at the point where the chamber was situated. After 200 monitor units, a charge of 64.5 nC was measured at 22 °C and 100.2 kPa. Estimate the effective volume of the chamber. The monitor is assumed to be independent of air pressure and temperature. The electron $s_{w,air}$ is 1.080 at the chamber position.
*Answer: 1 cm$^3$.*

**Solution:**
From the Bragg–Gray principle, the absorbed dose is given by $D_w = D_{air} s_{w,air}$. On the other hand, for an ion chamber measurement, the dose is given by the usual relation

$$D_{air} = \frac{q}{m_{air}} \frac{W_{air}}{e}$$

From the two expressions, we get:

$$m_{air} = \frac{q}{D_w} s_{w,air} \frac{W_{air}}{e}$$

that is,

$$volume = \frac{1}{\rho_{air}} \frac{q}{D_w} s_{w,air} \frac{W_{air}}{e}$$

Because the air density at 20 °C and 101.325 kPa will be used in this equation, the charge $q$ needs to be corrected for pressure and temperature, that is,

$$k_T = \frac{273.15 + T(°C)}{273.15 + T_{ref}(°C)} = \frac{273.15 + 22}{273.15 + 20} = 1.0068$$

$$k_P = \frac{P_{ref}}{P} = \frac{101.325}{100.2} = 1.0112$$

and be divided by 2 to get its value for 100 m.u. This yields

$$q = \frac{q'}{2} \times k_T \times k_P = 32.83 \text{ nC for 100 m.u.}$$

Therefore, the volume is

$$volume = \frac{1}{1.2045 \text{ kg m}^{-3}} \times \frac{32.83 \times 10^{-9} C}{1.0 \text{ Gy}} \times 1.080 \times 33.97 \text{ J } C^{-1}$$
$$= 1.0 \times 10^{-6} \text{ m}^3 = 1 \text{ cm}^3$$

5  A water calorimeter is irradiated with $^{60}$Co $\gamma$ rays and a temperature rise of $7.58 \times 10^{-4}$ K is measured at a reference point after an irradiation time of 300 s. A 1 cm$^3$ ionization chamber is then positioned with its center at the reference point. Determine the ionization current in the chamber if $T = 22$ °C and $P = 100.2$ kPa. For $^{60}$Co , $s_{w,air} = 1.1333$; the specific heat capacity of water is 4.186 kJ kg$^{-1}$K$^{-1}$. Assume all the correction factors involved to be equal to one.
*Answer: 325 pA.*

**Solution:**
The absorbed dose rate in a water calorimeter can be written as (see Eq. (11.18) in the textbook)

$$\dot{D}_w = \frac{c_w \Delta T}{t} \prod_i k_i$$

where $c_w$ is the specific heat capacity of water, $\Delta T$ is the temperature rise, $t$ is the measuring time, and $k_i$ are the necessary calorimeter correction factors, here taken as unity.
For an ionization chamber,

$$\dot{D}_w = \dot{D}_{air} s_{w,air} = \frac{i}{m_{air}} \frac{W_{air}}{e} s_{w,air}$$

From the two expressions results:

$$i = \frac{m_{air}}{W_{air}/e\, s_{w,air}} \frac{c_w \Delta T}{t}$$

where
- $m_{air} = 1.0\ cm^3 \times 1.2045 \times 10^{-3}\ g\ cm^{-3} = 1.2045 \times 10^{-3}\ g = 1.2045 \times 10^{-6}\ kg$
- $W_{air}/e = 33.97\ J\,C^{-1}$
- $s_{w,air} = 1.1333$
- $c_w = 4.186 \times 10^3\ J\ kg^{-1}K^{-1}$
- $\Delta T = 7.58 \times 10^{-4}K$
- $t = 300\ s$

These result in $i' = 3.309 \times 10^{-10}\ A = 330.9\ pA$, which needs to be corrected for pressure and temperature, that is,

$$k_T = \frac{273.15 + T(°C)}{273.15 + T_{ref}(°C)} = \frac{273.15 + 22}{273.15 + 20} = 1.0068$$

$$k_P = \frac{P_{ref}}{P} = \frac{101.325}{100.2} = 1.0112$$

Therefore the current is (note we are going from reference to actual conditions using factors designed to go from actual to reference conditions)

$$i = \frac{i'}{k_T \times k_P} = \frac{330.9\ pA}{1.0068 \times 1.0112} = 325\ pA$$

**6** The absorbed dose rate from a $^{60}Co$ γ-ray source was determined using a Fricke dosimeter immersed in a water phantom. After 50 h irradiation, an increase in absorbance at 304 nm of 0.364 was measured. Through calibration, the extinction coefficient $\epsilon$ (304 nm) was determined to be 220 m$^2$ mol$^{-1}$. The radiation chemical yield $G_{Fe3+}$ is $1.62 \times 10^{-6}$ mol kg$^{-1}$Gy$^{-1}$. Calculate the absorbed dose rate to water at the point where the Fricke dosimeter was placed. The optical path length of the solution is 10 mm, and its density 1.024 g cm$^{-3}$. The Fricke dosimeter may be considered large compared to the range of the electrons but small compared to the mean free path of the photons.
*Answer: 2 Gy h$^{-1}$.*

**Solution:**
The mean absorbed dose to a Fricke solution, $\bar{D}_F$, is expressed as (see Eq. (11.19) in the textbook)

$$\bar{D}_F = \frac{\Delta OD}{\rho_F\, \mathcal{L}\, \epsilon_F\, G_{Fe3+}}$$

where $\Delta OD$ is the measured change in optical density, $\rho_F$ is the solution density, $\mathcal{L}$ is the optical path length of the solution, $\epsilon_F$ is the extinction coefficient (difference in the molar linear absorption coefficient for the ferrous and ferric ions), and $G_{Fe3+}$ is the radiation chemical yield of the ferric ion.

The absorbed dose to water assuming CPE is given by

$$D_w = \bar{D}_F \, [\mu_{en}/\rho]_{w,Fricke}$$

For this case, we have
- $\Delta OD = 0.364$
- $\rho_F = 1.024 \times 10^3 \text{ kg m}^{-3}$
- $\mathcal{L} = 10 \text{ mm} = 0.01 \text{ m}$
- $\epsilon_F = 220 \text{ m}^2 \text{ mol}^{-1}$
- $G_{Fe3+} = 1.62 \times 10^{-6} \text{ mol kg}^{-1} \text{ Gy}^{-1}$

From the above, and dividing by the irradiation time of 50 h, results:

$$\dot{D}_F = \frac{0.364}{1.024 \times 10^3 \text{ kg m}^{-3} \times 0.01 \text{ m} \times 220 \text{ m}^2 \text{ mol}^{-1} \times 1.62 \times 10^6 \text{ mol/(kg Gy)}}$$
$$\times \frac{1}{50 \text{ h}} = 1.995 \text{ Gy h}^{-1}$$

From the Data Tables, for 1.25 MeV photons,

$$[\mu_{en}(k)/\rho]_w = 2.9598 \times 10^{-2} \text{cm}^2 \text{ g}^{-1}$$
$$\text{and} \quad [\mu_{en}(k)/\rho]_{Fricke} = 2.9514 \times 10^{-2} \text{cm}^2 \text{ g}^{-1}$$

Therefore

$$\dot{D}_w = 1.995 \text{ Gy h}^{-1}\text{h} \times \frac{0.029598}{0.029514} = 2.00 \text{ Gy h}^{-1}$$

7   Gamma rays of 35 keV from $^{125}$I are absorbed in a Fricke dosimeter solution to produce an average dose of 17 Gy. Assuming that 14.6 molecules of $Fe^{3+}$ are created per 100 eV, calculate the increase in optical density that would result at 304 mm in a 1 cm cell. What is the easiest way to increase $\Delta(OD)$ into a range where the accuracy of this method is optimal? Consider $\epsilon_F(304 \text{ nm}) = 2187 \, \text{l mol}^{-1} \text{ cm}^{-1}$.
*Answer: $\Delta(OD)=0.058$. Use a longer cell.*

**Solution:**
As in the previous exercise, the mean absorbed dose to a Fricke solution is expressed as (see Eq. (11.19) in the textbook)

$$\bar{D}_F = \frac{\Delta OD}{\rho_F \, \mathcal{L} \, \epsilon_F \, G_{Fe3+}}$$

and therefore $\Delta OD = \bar{D}_F \, \rho_F \, \mathcal{L} \, \epsilon_F \, G_{Fe3+}$, where
- $\bar{D}_F = 17$ Gy;
- $\rho_F = 1.024 \text{ g cm}^{-3} = 1.024 \times 10^{-3} \text{ kg cm}^{-3}$
- $\mathcal{L} = 1 \text{ cm}$
- $\epsilon_F = 2187 \text{ l mol}^{-1} \text{ cm}^{-1} = 2.187 \times 10^6 \text{cm}^2 \text{ mol}^{-1}$

- $G_{Fe3+}(35 \text{ keV}) = 14.6$ molecules per 100 eV, and as the conversion from molecules per joule to moles per joule is just Avogadro's constant, and $1 \text{ eV} = 1.6022 \times 10^{-19}$ J, we have:

$$G_{Fe3+} = \frac{\frac{14.6 \text{ molecules}}{6.022 \times 10^{23} \text{ molecules/mol}}}{100 \text{ eV} \times 1.6022 \times 10^{-19} \text{ J eV}^{-1}} = 1.5132 \times 10^{-6} \text{ mol J}^{-1}$$

Therefore

$$\Delta OD = 17 \text{ Gy} \times 1.024 \times 10^{-3} \text{ kg cm}^{-3} \times 1 \text{ cm} \times 2.187 \times 10^{6} \text{ cm}^2 \text{ mol}^{-1}$$
$$\times 1.5132 \times 10^{-6} \text{ mol J}^{-1}$$
$$= 0.058$$

For a given $\bar{D}_F$, to increase $\Delta OD$ where the only non-constant quantity is $\mathcal{L}$, use a longer cell.

Note that different units for $\epsilon(Fe^{3+})$ are often used in the literature, for example, in the present case:

$$\epsilon(Fe^{3+}) = 2187 \text{ M}^{-1} \text{ cm}^{-1} = 2187 \text{ (mol/liter)}^{-1} \text{ cm}^{-1}$$
$$= 2.187 \times 10^{6} \text{ mol}^{-1} \text{ cm}^2$$

**8** One liter of stirred Fricke solution is irradiated by a 1 MeV electron beam passing through an aperture 2 cm in diameter, for a period of 1 min. The solution is larger than the beam area and sufficient in thickness to stop the incident electrons. If $\Delta OD = 1.20$ at 304 nm in a 1 cm cell, what is the energy fluence rate at the aperture? Use the product recommended in ICRU Report 35 (ICRU, 1984a), $\epsilon_F \ G_{ICRU-35} = 352 \times 10^{-6} \text{ m}^2 \text{ kg}^{-1} \text{ Gy}^{-1}$. (Neglect backscattering.)

*Answer: $1.81 \times 10^4 \ Jm^{-2} s^{-1}$.*

**Solution:**
$\rho_F = 1.024 \text{ g/cm}^3 = 1024 \text{ kg/m}^3$
$\mathcal{L} = 1 \text{ cm} = 0.01 \text{ m}$
and

$$\bar{D}_F = \frac{\Delta OD}{\rho_F \ \mathcal{L} \ \epsilon_F \ G_{Fe3+}}$$

$$= \frac{1.20}{1024 \text{ kg m}^{-3} \times 0.01 \text{ m} \times 352 \times 10^{-6} \text{ m}^2 \text{ kg}^{-1} \text{ Gy}^{-1}} = 332.92 \text{ Gy}$$

As 1 l of solution weighs 1.024 kg,

$$\text{energy} = 332.92 \text{ Gy} \times 1.024 \text{ kg} = 340.91 \text{ J}$$
$$\text{aperture} = 3.1416 \times 10^{-4} \text{ m}^2$$

and the energy fluence rate is

$$\dot{\psi} = \frac{341 \text{ J}}{3.1416 \times 10^{-4} \text{ m}^2 \times 60 \text{ s}} = 1.809 \times 10^4 \text{ J m}^{-2} \text{ s}^{-1}$$

**9** An absorbed-dose calorimeter contains a 30 g core of graphite with a calibrating heater of resistance 10 $\Omega$. A current of 0.3 A is passed through it for 10 s under adiabatic conditions. A temperature increase of 0.42 K is measured. What is the value of the thermal capacity of the core material? What absorbed dose would cause a $\Delta T$ of 0.1 K under adiabatic conditions? *Answer: 714 J kg$^{-1}$ K$^{-1}$; 71.4 Gy.*

**Solution:**
The mean absorbed dose to the material 'm' of the calorimeter core is (see Eq. (11.14) in the textbook)

$$\bar{D}_{m} = c_{m}\Delta T$$

from where expressing the energy (joules) in terms of electrical quantities using $E = R\, i^2\, t$ we get

$$c_{m} = \frac{E}{\Delta T\, m} = \frac{10\,\Omega \times (0.3\text{A})^2 \times 10\text{ s}}{0.42\text{ K} \times 0.03\text{ kg}} = 714\,\frac{\text{J}}{\text{kg K}}$$

Thus

$$\bar{D}_{m} = c_{m}\,\Delta T = 714\,\frac{\text{J}}{\text{kg K}} \times 0.1\text{ K} = 71.4\text{ Gy}$$

# 12

# Ionization Chambers

**1** Determine the effective atomic number of water with respect to the photo-electric effect.
*Answer: 7.51.*

**Solution:**
The effective atomic number for the photoelectric effect is given by Eqs. (12.1) and (12.2) in the textbook, that is,

$$Z_{\mathrm{eff}}^{\mathrm{m}} = \sum_i a_i\, Z_i^{\mathrm{m}}, \quad \text{with m} \approx 3.5$$

and

$$a_i = \frac{\omega_i Z_i / A_i}{\sum\limits_i (\omega_i Z_i / A_i)}$$

where $\omega_i$ is the fraction by weight of element $i$ present in the mixture. For $H_2O$, we have $Z_H = 1$, $A_H = 1.008$, $Z_O = 8$, and $A_O = 16$, so that

$$\omega_H = \frac{1.008 \times 2}{16 + 1.008 \times 2} = 0.1119, \quad \omega_O = \frac{16}{16 + 1.008 \times 2} = 0.8881$$

$$a_H = \frac{0.11101}{0.55506} = 0.2, \qquad a_O = \frac{0.44405}{0.55506} = 0.8$$

and

$$Z_{\mathrm{eff}} = \sqrt[3.5]{0.2 \times 1^{3.5} + 0.8 \times 8^{3.5}} = 7.51$$

Note that, because $\omega_i$ can also be expressed in terms of $A_i$, the expressions for $a_H$ and $a_O$ reduce to

$$a_H = \frac{Z_H}{\sum Z} = \frac{1+1}{1+1+8} = 0.2, \qquad a_O = \frac{Z_O}{\sum Z} = \frac{8}{1+1+8} = 0.8$$

where the summation is not over the elements $i$ but over the constituents of the $H_2O$ molecule, hence H appears twice.

**2** The capacitance of a parallel-plate ionization chamber can be obtained from

$$C = \Delta q / \Delta V = 8.85 \times 10^{-14}\, a/d$$

*Fundamentals of Ionizing Radiation Dosimetry: Solutions to Exercises*, First Edition.
Pedro Andreo, David T. Burns, Alan E. Nahum, and Jan Seuntjens.
© 2017 Wiley-VCH Verlag GmbH & Co. KGaA. Published 2017 by Wiley-VCH Verlag GmbH & Co. KGaA.

where $\Delta q$ is the charge measured (in coulombs) as a result of a voltage change $\Delta V$ (in volts), $a$ is the collecting area in cm$^2$, and $d$ the plate separation in cm. The capacitance $C$ is in farads, and the numerical constant has units of F cm$^{-1}$. What are the capacitance and the collecting volume of a guarded parallel-plate chamber having a circular collecting region 2.5 cm in diameter, if an applied potential of 300 V is found to induce a charge of $5.21 \times 10^{-10}$ C? *Answer: 1.74 pF, 1.23 cm$^3$.*

**Solution:**
From the charge and voltage changes,
$\Delta q = 5.21 \times 10^{-10}$ C
$\Delta V = 300$ V
the capacitance is

$$C = \frac{\Delta q}{\Delta V} = \frac{5.21 \times 10^{-10} \text{ C}}{300 \text{ V}} = 1.737 \times 10^{-12} \text{ F} = 1.737 \text{ pF}$$

Together with the collecting area, $a = \pi\,(1.25)^2 = 4.91$ cm$^2$, the capacitance yields a plate separation

$$d = 8.85 \times 10^{-14} \frac{a}{C} = 8.85 \times 10^{-14} \text{ F cm}^{-1} \times \frac{4.91 \text{ cm}^2}{1.737 \times 10^{-12} \text{ F}} = 0.250 \text{ cm}$$

and the volume is

$$\text{volume} = a \times d = 4.91 \text{ cm}^2 \times 0.250 \text{ cm} = 1.23 \text{ cm}^3$$

**3**  Assume that a charge of $8.65 \times 10^{-10}$ C is measured with a cavity ionization chamber filled with air at 86.66 kPa, 23.5°C and 45% RH. (a) For the same irradiation, what charge would be measured if the chamber were filled with dry air at 101.325 kPa and 20°C? (b) What is the absorbed dose to dry air if the chamber volume is 3 cm$^3$ and recombination is negligible? *Answer: (a) 1.021 nC; (b) 9.60 mGy.*

**Solution:**
(a) The charge needs to be corrected for $P$, $T$, and humidity in the first case. From Figure 12.6 in the textbook, for 45% RH the ratio $i_h/i_a = 1.0027$ is the reciprocal of the humidity correction factor, $k_h$; therefore $k_h = 1/1.0027 = 0.9973$. Combining $k_h$ with the pressure and temperature correction results in

$$q = q' \times \frac{101.325}{P} \times \frac{273.15 + T}{273.15 + 20 \text{ °C}} \times k_h$$

$$= 8.65 \times 10^{-10} \times \frac{101.325}{86.66} \times \frac{273.15 + 23.5}{273.15 + 20} \times 0.9973$$

$$= 1.021 \times 10^{-9} \text{ C} = 1.021 \text{ nC}$$

(b) Knowing the charge that would be measured in dry air at 101.325 kPa and 20°C (for this air volume of 3 cm$^3$), we now just need to know the density of this air and the charge-to-energy conversion $W_{air}/e$ for dry air and use the standard equation

$$D_{air} = \frac{q}{v\,\rho_{air}}\frac{W_{air}}{e}$$

$$= \frac{1.021 \times 10^{-9}\ \text{C}}{3 \times 10^{-6}\ \text{m}^3 \times 1.2045\ \text{kg m}^{-3}} \times 33.97\ \frac{\text{J}}{\text{C}}$$

$$= 9.60 \times 10^{-3}\ \text{Gy} = 9.60\ \text{mGy}$$

Note that evaluation of the dose to humid air is less straightforward, for two reasons. Firstly, we would need to use the density of humid air for the $T$, $P$, and RH in question, which is derived from the complex equation given in the Appendix to Chapter 12. Secondly, we would need to use the $W_{air}/e$ value for this humid air, which differs from that for dry air ($33.97\,\text{J C}^{-1}$). This is not simply a case of applying the humidity correction $k_h = 0.9973$ because this correction includes not only the different $W_{air}/e$ value for humid air but also the different stopping power (which has to be excluded because it is implicitly included in the measured charge). This is also discussed in the Appendix to Chapter 12.

**4** A cylindrical ionization chamber with air-equivalent plastic walls has a collecting electrode of 3 mm diameter, an inside wall of 1.5 cm diameter, and a volume of 5 cm$^3$. It is filled with air and operates with a collecting potential of 100 V. The chamber is irradiated at 1.0 Gy min$^{-1}$ in a continuous x-ray beam. (a) Calculate the saturation current, the charge-collection efficiency, and the recombination correction factor. (b) Recalculate the recombination correction if the potential is tripled. (c) From answers (a) and (b) estimate the current that would be measured at each potential, use the two-voltage technique to re-estimate the saturation current, and compare it with the value obtained at (a).
*Answer: (a) 2.9548 nA, 0.9925, and 1.0075; (b) 1.0008; (c) 2.9328 nA and 2.9523 nA, the two-voltage technique reproduces the saturation current determined in (a).*

**Solution:**

(a) For continuous radiation, the collection efficiency is given by (see Eqs (12.24) and (12.25) in the textbook):

$$f = \frac{1}{1 + \xi^2}$$

where

$$\xi^2 = \mu_c \left(\frac{S^2}{V}\right)^2 \frac{q_0}{v\,t} = \mu_c \left(\frac{S^2}{V}\right)^2 \frac{\dot{q}_0}{v}$$

with

- $\mu_c$      constant, $6.73 \times 10^{13}\ \text{V}^2\,\text{m}^{-1}\,\text{C}^{-1}\,\text{s}$
- $S$      geometry factor
- $V$      polarizing potential
- $\dot{q}_0/v$      charge-rate (current) density.

We will need the geometry factor for a cylindrical chamber, which is given by $S = k_{cyl}(a - b)$, where $a$ is the inside wall radius (7.5 mm), $b$ the collector radius (1.5 mm), and

$$k_{cyl}^2 = \frac{1}{2}\left(\frac{a+b}{a-b}\right)\ln\left(\frac{a}{b}\right) = \frac{1}{2}\left(\frac{7.5+1.5}{7.5-1.5}\right)\ln\left(\frac{7.5}{1.5}\right) = 1.2071$$

Therefore $k_{cyl} = 1.0987$ and

$$S = 1.0987 \times (7.5 - 1.5) = 6.5920 \text{ mm} = 0.006592 \text{ m}.$$

Note that we convert to meters to maintain consistency with the constant $\mu_c$. This is also done in the following.

We need to determine now the charge-rate (current) density $\dot{q}_0/v$ taking into account that the dose rate is 1.0 Gy min$^{-1}$ and the chamber volume is 5 cm$^3$. Using the basic expression for the absorbed dose as a function of the air mass (v $\rho_{air}$) in an ionization chamber,

$$\dot{q}_0 = \frac{\dot{D}_{air}}{W_{air}/e}(v\ \rho_{air})$$

$$= \frac{1.0 \text{ Gy/min}}{33.97 \text{ J/C}} \times (5 \times 10^{-6} \text{ m}^3 \times 1.2045 \text{ kg m}^{-3})$$

$$= 1.7729 \times 10^{-7} \text{ C/min} = 2.9548 \times 10^{-9} \text{ C/s} = 2.9548 \text{ nA}$$

where $\rho_{air} = 1.2045$ kg m$^{-3}$ has been used. This is the saturation current (that would be measured if no recombination takes place).

The charge-rate density is

$$\frac{\dot{q}_0}{v} = \frac{2.9548 \times 10^{-9} \text{ C/s}}{5 \times 10^{-6} \text{ m}^3} = 5.9096 \times 10^{-4} \text{ C s}^{-1}\text{m}^{-3}$$

From the values above, we get for the operating potential

$$\xi^2(100 \text{ V}) = 6.73 \times 10^{13}\frac{V^2\ s}{C\ m} \times \frac{0.006592^4\ m^4}{100^2\ V^2}$$

$$\times 5.9096 \times 10^{-4}\frac{C}{s\ m^3} = 7.5102 \times 10^{-3}$$

(note the cancelation of units). Therefore the collection efficiency is

$$f(100 \text{ V}) = \frac{1}{1 + 7.5102 \times 10^{-3}} = 0.9925$$

and the corresponding recombination correction factor

$$k_s(100 \text{ V}) = \frac{1}{f} = 1.0075$$

(b) We only need to recalculate $\xi^2$ for 300 V, that is,

$$\xi^2(300 \text{ V}) = \xi^2(100 \text{ V}) \times \frac{100^2}{300^2} = 8.3447 \times 10^{-4}$$

and then get $f(300 \text{ V}) = 0.9992$ and $k_s(300 \text{ V}) = 1.0008$, that is, if the potential is multiplied by $n$ (=3), recombination is reduced by a factor of $n^2$ (=9).

(c) We use the saturation current calculated in (a) and the two recombination corrections to predict the measured currents for the two potentials:

at $V_1 = 100$ V, $\dot{q}_1 = \dot{q}_0 / k_{s,1} = 2.9548/1.0075 = 2.9328$ nA

at $V_2 = 300$ V, $\dot{q}_2 = \dot{q}_0 / k_{s,2} = 2.9548/1.0008 = 2.9523$ nA

Using these two 'measured' currents, we use the two-voltage technique, Eq. (12.28) in the textbook, to make a new estimate of the recombination correction at 300 V and hence the saturation current.

$$f_2 = \frac{n^2 - (q_2/q_1)}{n^2 - 1} = \frac{3^2 - (2.9523/2.9328)}{3^2 - 1} = 0.9992$$

$$k_{s,2} = \frac{1}{f_2} = 1.0008$$

and

$$\dot{q}_0 = \dot{q}_2 \, k_{s,2} = 2.9523 \times 1.0008 = 2.9548 \text{ nA}$$

This value coincides with the initial estimate of the saturation current, which is as it should be because the two-voltage technique uses the same underlying theory and approximations as used to evaluate the recombination corrections.

5  An ionization chamber is exposed to pulsed radiation. A charge of $7.05 \times 10^{-8}$ C is collected with an applied potential of 200 V and $5.51 \times 10^{-8}$ C at 90 V. Calculate the charge produced in the chamber. (*Hint*: You may solve for $u_1$ graphically or numerically.)

*Answer: 96.1 nC.*

**Solution:**

From Eq. (12.21) in the textbook, for pulsed radiation we have

$$\frac{q_{1,p}}{q_{2,p}} = n \, \frac{\ln(1 + u_1)}{\ln(1 + n\,u_1)} = \frac{V_1}{V_2} \frac{\ln(1 + u_1)}{\ln\left(1 + \frac{V_1}{V_2} u_1\right)}$$

$$\frac{7.05 \times 10^{-8}}{5.51 \times 10^{-8}} = 1.2795 = \frac{200}{90} \frac{\ln(1 + u_1)}{\ln\left(1 + \frac{200}{90} u_1\right)} = 2.222 \frac{\ln(1 + u_1)}{\ln(1 + 2.222\,u_1)}$$

that is,

$$\frac{\ln(1 + u_1)}{\ln(1 + 2.222\,u_1)} = 0.5758$$

which can be solved by trial and error:

| $u_1$ | $\ln(1 + u_1)/0.5758 \times \ln(1 + 2.222\,u_1)$ |
|---|---|
| 1 | 1.0289 |
| 0.5 | 0.9424 |
| 0.80 | 0.9992 |
| 0.81 | 1.0008 |
| 0.805 | 1.00 |

where the last value is obtained by interpolation between 0.80 and 0.81. Using Eq. (12.18), we then obtain

$$f_1 = \frac{q_{1,p}}{q_{0,p}} = \frac{1}{u_1} \ln(1 + u_1) = \frac{1}{0.805} \ln(1 + 0.805) = 0.7336$$

and therefore the charge produced in the chamber is

$$q_{0,p} = \frac{q_{1,p}}{0.7336} = \frac{7.05 \times 10^{-8} \text{ C}}{0.7337} = 96.10 \text{ nC}$$

**6** An ion chamber is irradiated within a phantom. A charge of $6.65 \times 10^{-9}$ C is collected at the operating potential, and a charge of $6.55 \times 10^{-9}$ C when the potential is halved. Calculate the charge produced in the chamber assuming that only initial recombination takes place.
*Answer: 6.753 nC.*

**Solution:**
We are given the potential ratio $n = V_1/V_2 = 2$, but no information is given on the potential values, nor information provided on the type of beam, continuous or pulsed, as initial recombination will be the same in either case. The general Eq. (12.12) in the textbook can be used in this case:

$$f = \frac{n - (q_1/q_2)}{n - 1} = \frac{2 - (6.65 \text{ nC}/6.55 \text{ nC})}{2 - 1} = 0.9847$$

yielding

$$k_s = 1/f = 1.0155$$

and therefore

$$q_0 = 6.65 \text{ nC} \times k_s = 6.753 \text{ nC}$$

**7** A thin-walled cylindrical ionization chamber has an internal diameter of 0.75 cm, a length of 1 cm, and a central electrode with a diameter of 0.1 cm. It is irradiated for 1 min in an accelerator that delivers a dose rate of 2 Gy min$^{-1}$ in air at the isocenter, at a pulse frequency of 200 Hz. Two measurements are made, at 400 V and 80 V, yielding collected charges of 2.60 pC and 2.57 pC, respectively. (a) Calculate the saturation charge using the two-voltage technique. (b) On the basis of the result obtained, determine the charge-density per pulse and the subsequent collection efficiency at the two voltages, estimating the charge produced in each case; these should be compared with the charges measured.
*Answer: (a) 2.6078 pC; (b) 5.9096 $\mu C \, m^{-3}$, $f_1$=0.9969, $f_2$=0.9848.*

**Solution:**
(a) As in previous exercises, for pulsed radiation we have Eq. (12.21) in the textbook:

$$\frac{q_{1,p}}{q_{2,p}} = n \frac{\ln(1 + u_1)}{\ln(1 + n \, u_1)}$$

where $n = V_1/V_2 = 5$ and

$$\frac{q_{1,p}}{q_{2,p}} = \frac{2.60 \text{ pC}}{2.57 \text{ pC}} = 1.0117$$

that is,

$$1.0117 = 5 \frac{\ln(1 + u_1)}{\ln(1 + 5u_1)}$$

or

$$\frac{\ln(1 + u_1)}{\ln(1 + 5\,u_1)} = 0.2023$$

This expression can be solved either numerically, by trial and error as in Exercise 5, or by using Figure 12.8b in the textbook for $n = 5$ and a $q$-ratio of 1.0117. The former yields $f_1 = 0.9970$ or $k_s = 1.0030$. The saturated charge then results

$$q_0 = q_{1,p} \times 1.003 = 2.6078 \text{ pC}$$

(b) The $q_0$ above is the total charge liberated within the chamber air volume for a dose rate of 2 Gy min$^{-1}$ during 1 min, that is, 2 Gy. We need the charge-density per pulse, which can be obtained from

$$\frac{q_0(\text{per pulse})}{v} = \frac{D_{air}(\text{per pulse})}{W/e} \rho_{air}$$

where

$-$ $D_{air}(\text{per pulse}) = \frac{2 \text{ Gy}}{\text{time} \times \text{freq}} = \frac{2 \text{ Gy}}{60 \text{ s} \times 200 \text{ s}^{-1}} = 1.6667 \times 10^{-4} \text{ Gy}$

$-$ $W/e = 33.97 \text{ J C}^{-1}$

$-$ $\rho_{air} = 1.2045 \text{ kg m}^{-3}$

that is, the charge-density per pulse becomes

$$\frac{q_0(\text{per pulse})}{v} = \frac{1.6667 \times 10^{-4} \text{ Gy}}{33.97 \text{ J C}^{-1}} \times 1.2045 \text{ kg m}^{-3} = 5.9096 \text{ μC m}^{-3}$$

Using Eqs. (12.15) and (12.16) in the textbook, that is,

$$f = \frac{\ln(1 + u)}{u}$$

and

$$u = \mu_p \frac{S^2}{V} \frac{q_0}{v}$$

where

$-$ $\mu_p$     constant, $3.02 \times 10^{10} \text{ V m C}^{-1}$

$-$ $S$     geometry factor

$-$ $V$     polarizing potential

$-$ $q_0/v$     charge-density per pulse.

we determine $u$ and $f$ for each voltage, and with the latter, the corresponding $q$ using the result for $q_0$ from (a).

The geometry factor for a cylindrical chamber is given by $S = k_{cyl}(a - b)$, where $a$ is the chamber inner radius (3.75 mm), $b$ the central electrode radius (0.5 mm), and

$$k_{cyl}^2 = \frac{1}{2}\left(\frac{a+b}{a-b}\right)\ln\left(\frac{a}{b}\right) = \frac{1}{2}\left(\frac{3.75+0.5}{3.75-0.5}\right)\ln\left(\frac{3.75}{0.5}\right) = 1.3174$$

Therefore $k_{cyl} = 1.1478$ and

$$S = 1.1478 \times (3.75 - 0.5) = 0.00373 \text{ m}$$

resulting in

$$u(V_1) = 0.00621; \ f(V_1) = 0.9969; \ q_1(V_1) = 2.6078 \text{ pC} \times 0.9969 = 2.60 \text{ pC}$$
$$u(V_2) = 0.03104; \ f(V_2) = 0.9848; \ q_2(V_2) = 2.6078 \text{ pC} \times 0.9848 = 2.57 \text{ pC}$$

that coincide with the charges measured.

# 13

# Chemical Dosimeters

**1** What are the dosimeters of choice for (a) determining the response saturation of another dosimeter at its high-dose-rate limit, (b) measuring the average absorbed dose in an irregular volume, and (c) mapping the relative dose in an electron beam in a steep-gradient region?
*Answer: (a) Calorimetric dosimeter; (b) liquid chemical dosimeter; (c) photographic or chemical film dosimeter.*

**2** A spectrophotometer yields an optical density of 1.2 for a Fricke solution. (a) What is the percentage of light transmitted through the Fricke solution? (b) Calculate the concentration of ferric ions, using data in the chapter of the textbook. Assume that the readout temperature is 25 °C.
*Answer: 6.3%; 0.552 mM.*

**Solution:**
(a) The relation between the optical density (OD) and the light transmission is given by

$$OD = \log_{10}\frac{I_0}{I}$$

where $I_0$ is the intensity of the incoming light and $I$ the intensity of the transmitted light. The percentage transmission is thus

$$100 \times \frac{I}{I_0} = 100 \times 10^{-OD} = 100 \times 10^{-1.2} = 6.3\%$$

(b) The molar extinction coefficient for species $x$ is defined as the optical density per unit concentration of species $x$ and per unit length of the light path of the read-out cuvette in the spectrophotometer. The concentration $[Fe^{3+}]$ is thus related to OD through

$$[Fe^{3+}] = \frac{OD}{\epsilon(Fe^{3+})\,\mathcal{L}}$$

$$= \frac{1.2}{2.174 \times 10^6 \ \text{mol}^{-1} \ \text{cm}^2 \times 0.001 \ \text{l cm}^{-3} \times 1 \ \text{cm}}$$

*Fundamentals of Ionizing Radiation Dosimetry: Solutions to Exercises*, First Edition.
Pedro Andreo, David T. Burns, Alan E. Nahum, and Jan Seuntjens.
© 2017 Wiley-VCH Verlag GmbH & Co. KGaA. Published 2017 by Wiley-VCH Verlag GmbH & Co. KGaA.

$$= 0.000552 \text{ mol l}^{-1}$$

$$= 0.552 \text{ mM}$$

Note that Section 13.4.1 in the textbook gives $\epsilon(Fe^{3+}) = 2174 \text{ M}^{-1} \text{ cm}^{-1}$, that is,

$$\epsilon(Fe^{3+}) = 2174 \text{ (mol/l)}^{-1} \text{ cm}^{-1} = 2174 \text{ mol}^{-1} \text{ l cm}^{-1}$$

$$= 2.174 \times 10^6 \text{ mol}^{-1} \text{ l cm}^{-1}$$

In the literature, one can find values of $\epsilon(Fe^{3+})$ expressed in any of these units; using for example the value "2174 mol$^{-1}$ l cm$^{-1}$" would avoid the factor "0.001 l cm$^{-3}$" in the denominator.

**3**  $\gamma$ rays from $^{192}$Ir are absorbed in a Fricke dosimeter solution resulting in an average dose of 17 Gy. The irradiation was conducted at a temperature of 21.5 °C. Data for $^{192}$Ir (from PIRS 1980, 2014) indicate a G(Fe$^{3+}$) of 1.589 $\pm$ 0.009 µmol J$^{-1}$. What increase in optical density, $\Delta$OD, measured at 304 nm at a temperature of 23 °C would result in a 1 cm long spectrophotometer cell? What is the easiest way to increase $\Delta$OD towards a range where the accuracy of this method is optimal?

*Answer: $\Delta OD = 0.059$. One way of increasing the signal is to use a cuvette with longer optical path length $\mathcal{L}$, for example, 5 cm instead of 1 cm.*

**Solution:**
The absorbed dose to the Fricke solution is given by

$$D_F = \frac{\Delta OD_{25,25}}{\rho_F \, \epsilon_F \, G \, \mathcal{L}}$$

or

$$\Delta OD_{25,25} = D_F \rho_F \, \epsilon_F \, G \, \mathcal{L}$$

where $\Delta OD_{25,25}$ represents the change in optical density at irradiation and readout temperature of 25 °C, $\rho_F$ is the density of the Fricke solution (1.0227 g cm$^{-3}$), $\epsilon_F$ is the molar extinction coefficient (=2.174 $\times$ 10$^6$ cm$^2$ mol$^{-1}$ for a readout temperature of 25 °C), $G$ is the radiation chemical yield of ferric relative to ferrous ions (1.589 $\pm$ 0.009 µmol J$^{-1}$ for $^{192}$Ir), and $\mathcal{L}$ is the length of the light path through the readout cell.

The optical density change at irradiation and readout temperature of 25 °C is related to the optical density change at the actual temperatures ($T_{irr}$, $T_{read}$) by:

$$\Delta OD_{25,25} = \Delta OD \, [1 + 0.0012(25 - T_{irr})][1 + 0.0069(25 - T_{read})]$$

In this case, therefore, the measured $\Delta$OD is given by

$$\Delta OD = \frac{\Delta OD_{25,25}}{[1 + 0.0012(25 - T_{irr})][1 + 0.0069(25 - T_{read})]}$$

$$= \frac{D_F \rho_F \, \epsilon \, G \, \mathcal{L}}{[1 + 0.0012(25 - T_{irr})][1 + 0.0069(25 - T_{read})]}$$

Hence,

$$\Delta OD = \frac{\begin{array}{c} 17 \text{ Gy} \times 1.0227 \times 10^{-3} \text{ kg cm}^{-3} \times 2.174 \times 10^{6} \text{ mol}^{-1} \text{ cm}^{2} \\ \times 1.589 \times 10^{-6} \text{ mol J}^{-1} \times 1 \text{ cm} \end{array}}{[1 + 0.0012(25 - 21.5)][1 + 0.0069(25 - 23)]}$$

$$= \frac{0.06006}{1.0042 \times 1.0138} = 0.0590$$

An optical density change of 0.0590 is a relatively low value to determine accurately. One way of increasing the signal is to use cuvettes with longer optical path length $\mathcal{L}$, for example, 5 cm instead of 1 cm. The measured signal is proportional to the optical path length, as can be seen in the previous equations.

**4**  One liter of Fricke solution at 25 °C is irradiated by a 1 MeV electron beam collimated to an aperture 2 cm in radius, for a period of 1 min. If $\Delta OD = 0.125$ at 303 nm in a 1.5 cm long cell, what was the energy-fluence rate at the aperture? Assume an $\epsilon_F G$ of 3.517 cm$^2$ J$^{-1}$ and neglect backscattering. *Answer: 471.4 J m$^{-2}$ s$^{-1}$.*

**Solution:**
The absorbed dose to the Fricke solution is

$$\bar{D}_F = \frac{\Delta OD}{\rho_F \, \epsilon_F \, G(Fe^{3+}) \, \mathcal{L}}$$

$$= \frac{0.125}{1.0227 \times 10^{-3} \text{ kg cm}^{-3} \times 3.517 \text{ cm}^2 \text{ J}^{-1} \times 1 \text{ cm}} = 34.75 \text{ Gy}$$

One liter of 25 °C solution weighs 1.0227 kg. The total energy is then

$$E = 34.75 \text{ Gy} \times 1.0227 \text{ kg} = 35.54 \text{ J}$$

For the aperture of $\pi \times 4 \times 10^{-4}$ m$^2$, the energy-fluence rate thus becomes

$$\dot{\psi} = \frac{35.54 \text{ J}}{\pi \times 4 \times 10^{-4} \text{ m}^2 \times 60 \text{ s}} = 471.36 \text{ J m}^{-2} \text{ s}^{-1}$$

**5**  What magnitude of magnetic field would be required to develop an EPR spectrometer using S-band microwave technology? *Answer: 0.1 T.*

**Solution:**
S-band microwaves are in the frequency range of 2–4 GHz. Assuming 3 GHz, the corresponding photon energy is (see Eq. (1.1) in the textbook)

$$k = h \nu = 4.136 \times 10^{-15} \text{ eV s} \times 3 \times 10^{9} \text{ s}^{-1} = 12.407 \times 10^{-6} \text{ eV}$$

The energy difference between the lowest energy state and the highest energy state is given by Eq. (13.29), that is,

$$\Delta E = g_e \, \mu_B \, B$$

where $g_e$ is the Landé factor (2.0023) and $\mu_B$ the Bohr magneton ($927.4 \times 10^{-26}$ J T$^{-1}$, see the Data Tables). In resonance, the external magnetic field required is (Eq. (13.30)):

$$B = \frac{k}{g_e\,\mu_B}$$
$$= \frac{12.407 \times 10^{-6}\ \text{eV} \times 1.6022 \times 10^{-19}\ \text{J eV}^{-1}}{2.0023 \times 927.4 \times 10^{-26}\text{J T}^{-1}}$$
$$\approx 0.1\ \text{T}$$

that is, a field strength of about one-third of that required for operation in the X-band (since the frequency is about one-third).

6 An absorbed dose of 1 cGy produces an optical density of 2 on a particular type of film. What is the film sensitivity (assuming linearity) in absorbance/Gy?
*Answer: 200 Gy$^{-1}$.*

**Solution:**
Film sensitivity is the absorbance change per amount of dose delivered. In this case, a dose of 1 cGy yielded an absorbance change of 2. The sensitivity is thus $2/0.01 = 200$ Gy$^{-1}$.

7 Calculate the number of silver atoms in a grain of AgBr used in radiographic films. The grain size is about 0.8 μm in diameter and the density of AgBr is $\rho_{\text{grain}} = 6.47$ g cm$^{-3}$.
*Answer: $5.55 \times 10^9$ atoms.*

**Solution:**
One mol of AgBr is 107.86 g + 79.90 g = 187.76 g and has $6.022 \times 10^{23}$ AgBr molecules. The mass of a single grain, assumed to be spherical, is given by

$$m_{\text{grain}} = \frac{4\pi r^3}{3}\rho_{\text{grain}}$$
$$= \frac{4}{3}\pi(0.4 \times 10^{-4})^3\ \text{cm}^3 \times 6.47\ \text{g cm}^{-3}$$
$$= 1.73 \times 10^{-12}\ \text{g}$$

In a single grain, there will thus be $1.73 \times 10^{-12}$ g of AgBr. The corresponding number of molecules of AgBr in this mass is

$$\# \text{ molecules of AgBr} = \frac{1.73 \times 10^{-12}\ \text{g}}{187.76\ \text{g}} \times 6.022 \times 10^{23}$$
$$= 5.55 \times 10^9\ \text{molecules}$$

There is one single atom of Ag per molecule of AgBr. Hence,

$$\# \text{ atoms of Ag} = 5.55 \times 10^9\ \text{atoms}$$

**8** A beam of 100 keV electrons is incident on a radiographic film and an optical density of 2.5 is measured. The grain diameter is 0.6 μm and, in the film, the effective number of grains per $cm^2$ is $10^{10}$. Calculate the dose giving rise to the OD measured.

*Answer: 28 mGy.*

**Solution:**

We use the relation between optical density OD and electron fluence, $\Phi_e$

$$OD = 0.4343 \, a^2 \, N_g \, \Phi_e$$

The electron fluence is then given by

$$\Phi_e = \frac{OD}{0.4343 \, a^2 \, N_g}$$

$$= \frac{2.5}{0.4343 \times (0.3 \times 10^{-4})^2 \, \pi \, cm^4 \times 10^{10} \, cm^{-2}}$$

$$= 7.2 \times 10^7 \, cm^{-2}$$

The dose corresponding to this electron fluence can be estimated using

$$D = \Phi_e \frac{S_{el}}{\rho}$$

Taking a mass stopping power of 2.401 MeV $cm^2$ $g^{-1}$ for 100 keV electrons in silver, we arrive at

$$D = 7.2 \times 10^7 \, cm^{-2} \times 2.401 \, MeV \, cm^2 \, g^{-1} \times 1.6 \times 10^{-10} \, Gy \, MeV^{-1} \, g$$

$$= 0.028 \, Gy = 28 \, mGy$$

# 14

## Solid-State Dosimeters

**1** Is the total light output from a TL dosimeter independent of the heating rate? If not, which effect limits it?
*Answer: No. Thermal quenching limits it.*

**2** The Randall–Wilkins theory relates the glow-peak temperature $T_m$ to the heating rate $\dot{h}_T$ and the energy depth of the trap $E_{trap}$ by

$$\frac{E_{trap}}{k_B\, T_m^2} = \frac{v}{\dot{h}_T}\, \exp\left(-E_{trap}/k_B T_m\right)$$

Show from this expression that the glow-peak temperature increases monotonically with the heating rate and plot $T_m$ as a function of $\dot{h}_T$ for $E_{trap} = 0.85$ eV and for $E_{trap} = 1$ eV. What is the glow-peak temperature for a heating rate of 40 °C min$^{-1}$?
*Answer: The glow-peak temperature is 138 °C.*

**Solution:**
The Randall–Wilkins equation is an implicit equation in $T_m$ and cannot be cast in explicit form. Therefore, a numerical method must be used. To this end, define the function

$$f(T_m) = \frac{E_{trap}}{k_B\, T_m^2} - \frac{v}{\dot{h}_T}\, \exp\left(-E_{trap}/k_B T_m\right)$$

For given values of the energy depth $E_{trap}$, $T_m$ can be found as a positive solution of $f(T_m) = 0$. Using, for example, *Mathematica* numerical root finding, the dependencies shown in Figure 14.1 are found for $E_{trap} = 0.85$ eV and $E_{trap} = 1$ eV, for $v = 10^9$ s$^{-1}$.

**3** Using the Randall–Wilkins expression in the previous exercise, show that for $v = 10^9$ s$^{-1}$ and $\dot{h}_T = 1$ K s$^{-1}$, the glow-peak temperature can be approximated by $T_m = 489$ K eV$^{-1} E_{trap}$.

**Solution:**
Using the same methodology as before, for given values of the heating rate, the functional dependence of $T_m$ versus $E_{trap}$ can be determined. The result

*Fundamentals of Ionizing Radiation Dosimetry: Solutions to Exercises,* First Edition.
Pedro Andreo, David T. Burns, Alan E. Nahum, and Jan Seuntjens.
© 2017 Wiley-VCH Verlag GmbH & Co. KGaA. Published 2017 by Wiley-VCH Verlag GmbH & Co. KGaA.

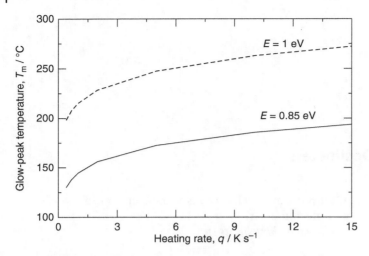

**Figure 14.1** Peak temperature $T_m$ as a function of the heating rate $\dot{h}_T$ for $E_{trap} = 0.85$ eV and $E_{trap} = 1$ eV.

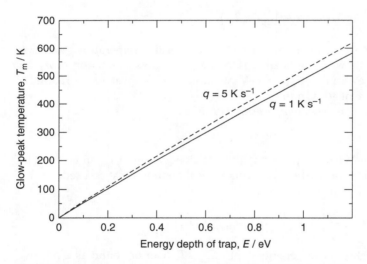

**Figure 14.2** Peak temperature $T_m$ as a function of $E_{trap}$ for heating rates $q = \dot{h}_T = 1$ K s$^{-1}$ and $q = 5$ K s$^{-1}$.

is shown in Figure 14.2, which shows a linear dependence of $T_m$ versus $E_{trap}$. The slope value for $q = 1$ K s$^{-1}$ amounts to 489 K eV$^{-1}$.

**4** A TLD has been calibrated in terms of air kerma in a beam of $^{60}$Co γ rays, and the calibration coefficient, in full build-up conditions, is 5.05 mGy nC$^{-1}$. The detector is used to measure air kerma in a 100 kV (HVL: 4.0 mm Al; effective energy: 50 keV) diagnostic radiology beam where the reading of the TLD amounts to 3.9 nC. Determine the air kerma in the 100 kV

diagnostic radiology beam taking into account the experimental results shown in Figure 14.7 of the textbook.
*Answer: 14.9 mGy.*

**Solution:**
The LiF TLD calibration coefficient in $^{60}$Co amounts to $N_K = 5.05$ mGy nC$^{-1}$. This air kerma calibration coefficient is not immediately applicable to the 100 kV diagnostic radiology beam as the LiF TLD shows a significant air kerma energy dependence, according to the results illustrated in Figure 14.7a of the textbook. On the basis of this graph, a correction to the calibration coefficient is required that takes into account that the detector response is higher by about 32% in the 100 kV (4.00 mm Al; effective energy: 50 keV) beam. Note that we are considering the measured response as a function of the effective energy to arrive at this value. Therefore,

$$N_K(100 \text{ kV}) = N_K(^{60}\text{Co}) \times \frac{1}{1.32} = \frac{5.05 \text{ mGy nC}^{-1}}{1.32} = 3.83 \text{ mGy nC}^{-1}$$

The air kerma measured by the LiF TLD in the diagnostic beam is thus

$$K_{\text{air}} = N_K(100 \text{ kV}) \times M_{\text{TLD}}(100 \text{ kV})$$
$$= 3.83 \text{ mGy nC}^{-1} \times 3.9 \text{ nC} = 14.9 \text{ mGy}$$

5   An OSL detector was used in a calibration experiment with the center of its sensitive volume placed at the depth of maximum dose in a water-equivalent plastic phantom in a 6 MV photon beam. A cylindrical 0.6 cm$^3$ ionization chamber with $N_{D,w,Q=6 \text{ MV}} = 4.857 \times 10^{-2}$ Gy nC$^{-1}$ was used to establish the absorbed dose to water at 10 cm depth, SSD 100 cm, for a field size 10 cm $\times$ 10 cm at the surface of the phantom. The percentage dose at the depth of 10 cm in the 6 MV beam for these conditions is 66.7%. Recorded chamber readings in nC, corrected for pressure and temperature, were 20.25, 20.30, 20.28, and 20.24; the ion recombination correction was to 1.0045 and the polarity correction 1.0013. A sequential set of OSLD readings (rdg) amounted to 351.04, 330.54, 342.41, and 328.67. A background reading by the same chip, taken before irradiation, was 19.23. Determine the OSLD chip calibration coefficient and discuss the measurement reproducibility.
*Answer: 4.65 × 10$^{-3}$ Gy rdg$^{-1}$ ± 1.6%. The type A standard uncertainties are 1.5% and 0.07% on the mean OSLD and the mean ionization chamber measurements, respectively.*

**Solution:**
The average chamber reading amounts to 20.27 nC, with a type A standard uncertainty of the mean of 0.07%. The average OSLD reading amounts to 338.2 with type A standard uncertainty of the mean equal to 1.56%.
We determine first the absorbed dose to water at 10 cm depth and at the depth of maximum dose, $z_{\text{max}}$:

$$D_w(z = 10 \text{ cm}) = \bar{M}_{\text{ch}} N_{D,w,Q=6 \text{ MV}} k_s k_{\text{pol}}$$
$$= 20.27 \text{ nC} \times 4.857 \times 10^{-2} \text{ Gy nC}^{-1} \times 1.0045 \times 1.0013$$
$$= 0.990 \pm 0.07\% \text{ Gy}$$

where it has been assumed that the uncertainty of $k_s$ and $k_{pol}$ are negligible, and

$$D_w(z_{max}) = \frac{D_w(z = 10 \text{ cm})}{\text{PDD}} = \frac{0.990}{0.667} = 1.484 \pm 0.07\% \text{ Gy}$$

The absorbed dose to water calibration coefficient of the OSL dosimeter in the 6 MV photon beam is obtained (subtracting the OSLD background from the mean OSDL reading) as

$$N_{D,w,Q=6\text{ MV}}^{\text{OSLD}} = \frac{D_w(z_{max})}{\bar{M}_{\text{OSLD}} - M_{\text{OSLD bckg}}} = \frac{1.484 \text{ Gy}}{338.2 - 19.23}$$
$$= 4.654 \times 10^{-3} \text{ Gy per reading}$$

and its type A uncertainty ($u_A$) is obtained combining in quadrature those of the numerator and the denominator, that is,

$$u_A[N_{D,w,Q=6\text{ MV}}^{\text{OSLD}}] = \sqrt{u_A^2[D_w(z_{max})] + u_A^2[\bar{M}_{\text{OSLD}}]} = \sqrt{0.07^2 + 1.56^2} = 1.56\%$$

Comparing the type A standard uncertainties of the respective mean values, it can be seen that the OSLD is significantly less reproducible than the ionization chamber.

**6**  Čerenkov radiation generated in the plastic fiber used as light guide in plastic scintillator detectors complicates the interpretation of the scintillation signal measured. The Čerenkov spectrum, given as the number of photons $N$ emitted by a charged particle with charge $z\,e$ per unit path length and per interval unit of photon energy, is (Tamm–Frank formula)

$$\frac{d^2N}{dE dx} = \frac{\alpha^2 z^2}{r_e mc^2} \left(1 - \frac{1}{\beta^2 n^2(E)}\right)$$

where $\beta$ is the ratio of the charged-particle velocity to the speed of light in vacuum, $n(E)$ is the refractive index of the fiber, $\alpha$ is the fine-structure constant, $r_e$ is the electron radius, and $m$ is the charged-particle mass. Demonstrate that the rate at which energy is lost by the charged particle per unit wavelength of photons $\lambda$ emitted is proportional to $1/\lambda^3$.

**Solution:**
From the Tamm–Frank formula,

$$\frac{d^2N}{dE\,dx} = \frac{\alpha^2 z^2}{r_e mc^2} \left(1 - \frac{1}{\beta^2 n^2(E)}\right)$$

we can find the energy loss in Čerenkov radiation of the charged particle per unit path length ($dE = \hbar\omega\, dN$):

$$\frac{dE}{dx} = \frac{(ze)^2}{4\pi\epsilon_0 c^2} \int_{v > \frac{c}{n}} \omega \left(1 - \frac{1}{\beta^2 n^2(\omega)}\right) d\omega$$

This integral is calculated over the frequencies $\omega$ for which the particle's speed $v$ is greater than the speed of light in the medium, $c/n(\omega)$. The integral converges (finite) because at high (x-ray) frequencies, the refractive index $n$ becomes less than unity, whereas for extremely high frequencies it becomes unity. We can recast this energy loss to be differential in, for example, $\omega$ to arrive at

$$\frac{\mathrm{d}E}{\mathrm{d}\omega} \propto \omega$$

Based on the relation, $\omega = 2\pi c / \lambda$, one arrives at the equivalent proportionality,

$$\frac{\mathrm{d}E}{\mathrm{d}\lambda} \propto \lambda^{-3}$$

**7** For the planning of an experiment using a diamond detector, preliminary measurements with a Farmer-type ionization chamber ($0.6$ cm$^3$) are performed at a reference depth of 5 cm in water in a 6 MV photon beam (field size 10 cm $\times$ 10 cm, SSD 100 cm), and the average charge recorded is 20.05 nC. Calculate the expected charge that will be generated in a diamond detector with a volume of 3.5 mm$^3$ placed with its effective point of measurement at the same position as the ionization chamber. *Note 1*: the $W/e$ values for diamond and air are 11.3 eV/electron–hole pair and 33.97 eV per ion pair, respectively, and $\rho_{\mathrm{diamond}} = 3.52$ g cm$^{-3}$. *Note 2*: Since this is an order-of-magnitude calculation, the difference in interaction coefficients between diamond and air may be ignored.

*Answer: The diamond signal collected over the same irradiation time (strictly, the same number of monitor units, MU) amounts to 1.02 μC.*

**Solution:**
The absorbed dose to the detector-sensitive volume for the ionization chamber and for the diamond can be assumed to be the same (see Note 2) and is denoted by $D_{\mathrm{det}}$. The absorbed dose to the detector volume of the ionization chamber amounts to

$$D_{\mathrm{det}} = \frac{W_{\mathrm{air}}}{e} \times \frac{q_{\mathrm{air}}}{m_{\mathrm{air}}} = \frac{W_{\mathrm{air}}}{e} \times \frac{q_{\mathrm{air}}}{\rho_{\mathrm{air}} v_{\mathrm{air}}}$$

$$= 33.97 \mathrm{J} \, \mathrm{C}^{-1} \times \frac{20.05 \times 10^{-9} \, \mathrm{C}}{1.205 \times 10^{-6} \, \mathrm{kg} \, \mathrm{cm}^{-3} \times 0.6 \, \mathrm{cm}^3} = 0.94 \, \mathrm{Gy}$$

Using this detector dose $D_{\mathrm{det}}$, the signal expected from the diamond detector is given by

$$q_{\mathrm{diamond}} = D_{\mathrm{det}} \frac{e}{W_{\mathrm{diamond}}} \rho_{\mathrm{diamond}} v_{\mathrm{diamond}}$$

$$= 0.94 \, \mathrm{Gy} \times \frac{1}{11.3 \, \mathrm{J} \, \mathrm{C}^{-1}} \times 3.52 \times 10^{-3} \, \mathrm{kg} \, \mathrm{cm}^{-3} \times 3.5 \times 10^{-3} \, \mathrm{cm}^3$$

$$= 1.02 \, \mu\mathrm{C}$$

**8** For a dose rate of 2 Gy min$^{-1}$ in a continuous radiation beam, a diamond detector yields 5 mA. What will the current be at a dose rate of 20 Gy min$^{-1}$ if the diamond has a sublinearity factor ($\Delta$) of 0.98?
*Answer: 47.7 μA.*

**Solution:**

The expression that describes the relative current as a function of dose rate is given by (Fowler, 1966)

$$i \approx \dot{D}^{\Delta}$$

The relative change in measured current in this case is thus given by

$$i_{20} = i_2 \frac{\dot{D}_{20}^{\Delta}}{\dot{D}_2^{\Delta}}$$

$$= 5\,\mu\text{A} \times \left( \frac{20\,\text{Gy min}^{-1}}{2\,\text{Gy min}^{-1}} \right)^{0.98}$$

$$= 47.7\,\mu\text{A}$$

This result is about 5% lower than expected assuming no dose-rate dependence (50 $\mu$A). This is due to the density effect in the electronic stopping power of diamond.

9  MOSFET calibrations were performed in a 6 MV photon beam in a water phantom at SSD = 100 cm, field size 10 cm × 10 cm and 5 cm depth. An absorbed dose of 1 Gy was delivered, determined with ionization chamber measurements. The MOSFET's threshold voltage $V_{th}$ values were 129 mV and 265 mV before and after irradiation, respectively. Determine the MOSFET absorbed dose to water calibration coefficient
*Answer: 7.35 mGy mV$^{-1}$.*

**Solution:**

The MOSFET threshold voltage change as a result of the delivered absorbed dose, $D_{ref}$, is given by

$$\Delta V_{th} = V_{th,\,post} - V_{th,pre}$$

Given that the MOSFET was irradiated at the position where the reference dose is known, the calibration coefficient in terms of absorbed dose to water is given by

$$N_{D,w}^{\text{MOSFET}} = \frac{D_{ref}}{\Delta V_{th}} = \frac{1\,\text{Gy}}{(265 - 129)\,\text{mV}} = 7.35 \times 10^{-3}\,\text{Gy/mV}$$

# 15

# Reference Dosimetry for External Beam Radiation Therapy

**1** The nominal cavity dimensions of a cylindrical ionization chamber are a radius of 3.2 mm and a length of 24 mm. (a) Determine its theoretical dose-to-air chamber coefficient, $N_{D,\text{air}}$, in $^{60}$Co $\gamma$ rays (ignoring the presence of a central collector). (b) Repeat the calculation of $N_{D,\text{air}}$ assuming that there would be an error of $-0.1$ mm in the radius during the manufacturing process (i.e., $r = 3.1$ mm). (c) For an air-kerma calibration coefficient of the chamber obtained in a standards laboratory of $N_K = 41.62$ mGy nC$^{-1}$, which yields a value of $N_{D,\text{air}} = 40.87$ mGy nC$^{-1}$, what would be the chamber radius, assuming its length coincides with the nominal value given here?
*Answer: (a) 36.53 mGy nC$^{-1}$; (b) 38.92 mGy nC$^{-1}$; (c) r = 3.0 mm.*

**Solution:**

(a) The relation between $N_{D,\text{air}}$ and the chamber air cavity volume $v$ is given by Eq. (15.18) in the textbook:

$$N_{D,\text{air},^{60}\text{Co}} = \frac{W_{\text{air}}/e}{v\,\rho_{\text{air}}} = \frac{W_{\text{air}}/e}{m_{\text{air}}}$$

where $\rho_{\text{air}} = 1.2045 \times 10^{-3}$ g cm$^{-3}$ and $W_{\text{air}}/e = 33.97$ J C$^{-1}$.

| | (a) | (b) |
|---|---|---|
| $r$ (cm) = | 0.32 | 0.31 |
| $l$ (cm) = | 2.40 | 2.40 |
| $v$ (cm$^3$) = | 0.772 | 0.725 |
| $m_{\text{air}}$ (kg) = | 9.300E−07 | 8.728E−07 |
| $N_{D,\text{air}}$ (Gy C$^{-1}$) = | 3.653E+07 | 3.892E+07 |
| $N_{D,\text{air}}$ (mGy nC$^{-1}$) = | 36.53 | 38.92 |

(b) A decrease of 0.1 mm in the radius would increase $N_{D,\text{air}}$ by nearly 7% for this chamber.

(c) Given the experimental $N_{D,\text{air}}$, we need to determine the chamber radius (assuming that its length is the same, $l = 24$ mm). Note that the experimental value is ~12% higher than the nominal $N_{D,\text{air}}$ obtained in (a), that is, the experimental $N_{D,\text{air}}$ would correspond to a chamber having 0.95 times the nominal radius:

*Fundamentals of Ionizing Radiation Dosimetry: Solutions to Exercises*, First Edition.
Pedro Andreo, David T. Burns, Alan E. Nahum, and Jan Seuntjens.
© 2017 Wiley-VCH Verlag GmbH & Co. KGaA. Published 2017 by Wiley-VCH Verlag GmbH & Co. KGaA.

$$N_{D,air} \, (\mathrm{mGy\,nC^{-1}}) = \qquad 40.873$$
$$N_{D,air} \, (\mathrm{Gy\,C^{-1}}) = \qquad 4.09E+07$$
$$m_{air} \, (\mathrm{kg}) = \qquad 8.31E{-}07$$
$$v \, (\mathrm{cm^3}) = \qquad 0.690$$
$$l \, (\mathrm{cm}) = \qquad 2.40$$
$$r \, (\mathrm{cm}) = \qquad 0.30$$

2   An air-kerma calibrated cylindrical ionization chamber has an $N_{D,air}$ dose-to-air chamber coefficient of 87.20 mGy nC$^{-1}$. Measurements made to calibrate a 6 MV photon beam (TPR$_{20,10} = 0.662$) at a reference depth of 5 cm in a water phantom with a field of 10 cm × 10 cm yield a collected charge of 26.90 nC at $P = 100$ kPa, $T = 22.9$ °C for an ion-collection efficiency of 0.992. Determine the absorbed dose to water at the reference depth. Additional data for the chamber are $p_{dis} = 0.994$, $p_{wall} = p_{cel} = 1.0$. For megavoltage photon beams, the water/air stopping-power ratio can be approximated by

$$s_{w,air} = 1.3443 - 1.2411 \times \mathrm{TPR}_{20,10} + 2.4977 \times \mathrm{TPR}^2_{20,10}$$
$$-1.7034 \times \mathrm{TPR}^3_{20,10}$$

*Answer: $D_w = 2.70$ Gy.*

**Solution:**
For the $N_K - N_{D,air}$ formalism, the absorbed dose to water at the reference depth is given by Eq. (15.19) in the textbook:

$$D_{w,Q}(z_{ref}) = M_{w,Q}(z_{ref}) \, N_{D,air} \, (s_{w,air})_Q \, p_{ch,Q}$$
$$= M_{w,Q}(z_{ref}) \, N_{D,air} \, (s_{w,air})_Q \, p_{dis} \, p_{wall} \, p_{cel}$$

where
– $N_{D,air} = 87.20$ mGy nC$^{-1}$
– $M_{w,Q}(z_{ref}) = q \, k_{TP} \, k_s$, with $q = 26.90$ nC
– $k_{TP} = \dfrac{273.15 + T\,°C}{273.15 + T_{ref}\,°C} \dfrac{P_{ref}}{P} = \dfrac{273.15 + 22.9}{273.15 + 20.0} \dfrac{101.325}{100.0} = 1.0233$
– $k_s = \dfrac{1}{0.992} = 1.0081$, that is, the saturation correction factor is the inverse of the ion-collection efficiency.

Therefore

$$M_{w,Q}(z_{ref}) = 26.90 \text{ nC} \times 1.0233 \times 1.0081 = 27.75 \text{ nC}$$

Using the analytical expression for the stopping-power ratio water-to-air, which for TPR$_{20,10} = 0.662$ yields $s_{w,air} = 1.1231$, and the various perturbation correction factors, the absorbed dose to water results:

$$D_{w,Q}(z_{ref}) = 27.75 \text{ nC} \times 87.20 \text{ mGy nC}^{-1} \times 1.1231 \times 0.994 \times 1.0 \times 1.0$$
$$= 2701 \text{ mGy} = 2.701 \text{ Gy}$$

**3** (a) Repeat the previous exercise for the case of a PMMA phantom (i.e., determine $D_{\text{PMMA}}$), assuming all the data remain as given there; the relevant stopping-power ratio should be estimated as a Bragg–Gray type, following the approximate procedure described in Chapter 9. (b) At what depth in PMMA should the chamber center be placed, equivalent to the 5 cm depth in water? (*Hint:* the number of electrons per g cm$^{-2}$ should be the same for both materials). (c) Assuming PCPE conditions, estimate the corresponding $D_w$ at 5 cm depth in water from the $D_{\text{PMMA}}$ determined in (a).
*Answer: (a) $D_{PMMA} = 2.63$ Gy; (b) $z_{PMMA} = 4.3$ cm; (c) $D_w = 2.71$ Gy.*

**Solution:**

(a) The absorbed dose at the reference depth is approximated by

$$D_{\text{PMMA}} = M_{\text{PMMA}} \; N_{D,\text{air}} \; s_{\text{PMMA,air}} \; p_{\text{ch},Q}$$

where it is assumed that only $s_{\text{PMMA,air}}$ differs from the previous exercise.[1] To estimate its value, we follow the approximate procedure described in Section 9.3.2.1 of the textbook for the Bragg–Gray stopping-power ratios:

– For a 6 MV photon beam, the equivalent monoenergetic photon energy is taken from Figure 9.11 as

$$\bar{k} = 0.36 \times \text{MV} = 2.16 \text{ MeV}$$

– Assuming that all the secondary electrons are liberated through Compton interactions, the mean initial Compton electron energy $\bar{E}_0$ is obtained from $\bar{k} \times \sigma_{\text{tr}}^{\text{KN}} / \sigma^{\text{KN}}$. From the photon Klein–Nishina cross sections in the electronic Data Tables,

$$\bar{E}_0 = \bar{k} \times \frac{\sigma_{\text{tr}}^{\text{KN}}}{\sigma^{\text{KN}}} = 2.16 \times \frac{7.542 \times 10^{-26}}{1.399 \times 10^{-25}} = 1.165 \text{ MeV}$$

– The mean energy in the equilibrium 'slowing-down spectrum' of the secondary electrons created by the Compton electrons is $\bar{E}_z \approx 0.5 \times \bar{E}_0 = 0.582$ MeV, and from the Data Tables,

$$s_{\text{PMMA,air}}^{\text{BG}} = \frac{[S_{\text{el}}(\bar{E}_z)/\rho]_{\text{PMMA}}}{[S_{\text{el}}(\bar{E}_z)/\rho]_{\text{air}}} = \frac{1.9142}{1.7517} = 1.0928$$

Therefore, using the data from the previous exercise together with the estimated PMMA stopping-power ratio, one gets

$$D_{\text{PMMA}} = 27.75 \text{ nC} \times 87.20 \text{ mGy nC}^{-1} \times 1.0928 \times 0.994 \times 1.0 \times 1.0$$
$$= 2628 \text{ mGy} = 2.628 \text{ Gy}$$

---

1 Note that physically it is not possible that the two readings be identical at the same depth. In the water phantom, the charge will be generated by interactions in water, whereas in PMMA it will be generated by interactions in PMMA, and the cross-section data and density differ in the two media. Assuming that $M_{\text{PMMA}} = M_{\text{w}}$ and $(p_{\text{ch},Q})_{\text{PMMA}} = (p_{\text{ch},Q})_{\text{w}}$ is, therefore, an approximation.

(b) If the number of electrons per g cm$^{-2}$ is constant

$$z_{\text{PMMA}} \, \rho_{\text{PMMA}} \left(\frac{Z}{A}\right)_{\text{PMMA}} N_A = z_{\text{w}} \, \rho_{\text{w}} \left(\frac{Z}{A}\right)_{\text{w}} N_A$$

and using $\rho$ and $Z/A$ values from the Data Tables,

$$
\begin{aligned}
z_{\text{PMMA}} &= \frac{\rho_{\text{w}}}{\rho_{\text{PMMA}}} \times \frac{(Z/A)_{\text{w}}}{(Z/A)_{\text{PMMA}}} \times z_{\text{w}} \\
&= \frac{0.9982}{1.19} \times \frac{0.555087}{0.539369} \times 5 \text{ cm} = 4.32 \text{ cm}
\end{aligned}
$$

(c) Under PCPE, the absorbed dose in the two media can be approximated by

$$D_{\text{w}} = \beta_{\text{w}} \, (K_{\text{el}})_{\text{w}} = \beta_{\text{w}} \Psi (\mu_{\text{en}}/\rho)_{\text{w}}$$
$$D_{\text{PMMA}} = \beta_{\text{PMMA}} \, (K_{\text{el}})_{\text{PMMA}} = \beta_{\text{PMMA}} \Psi (\mu_{\text{en}}/\rho)_{\text{PMMA}}$$

where $\beta_{\text{med}}$ is the ratio between absorbed dose and electronic (collision) kerma. From these equations, we obtain:

$$\frac{D_{\text{w}}}{D_{\text{PMMA}}} = \frac{\beta_{\text{w}}}{\beta_{\text{PMMA}}} \, (\mu_{\text{en}}/\rho)_{\text{w,PMMA}}$$

hence

$$D_{\text{w}} = D_{\text{PMMA}} \frac{\beta_{\text{w}}}{\beta_{\text{PMMA}}} \, (\mu_{\text{en}}/\rho)_{\text{w,PMMA}}$$

The $\beta_{\text{med}}$ ratio can be taken to be ~1 and for the $\mu_{\text{en}}/\rho$, ratio values are interpolated at the equivalent monoenergetic photon energy (2.16 MeV), that is, $(\mu_{\text{en}}/\rho)_{\text{w}} = 0.02543$ and $(\mu_{\text{en}}/\rho)_{\text{PMMA}} = 0.02464$ from the Data Tables, both in cm$^2$ g$^{-1}$.

The absorbed dose to water is therefore

$$D_{\text{w}} \approx 2.628 \text{ Gy} \times 1.0 \times \frac{0.02543}{0.02464} = 2.713 \text{ Gy}$$

which differs by 0.4% from the calculation of $D_{\text{w}}$ in the previous exercise (2.701 Gy).

4 A plane-parallel ionization chamber with 2 mm cavity height and 25 mm diameter is positioned with the inner surface of its front wall at a depth of 2 cm in a polystyrene phantom. An electron beam of 15 MeV incident on the phantom yields a charge reading of 64.2 nC at $P = 102$ kPa, $T = 25.2\,°$C at an ion-collection efficiency of 0.973. (a) What is the absorbed dose in the cavity air? (b) What is the absorbed dose in polystyrene at 2 cm depth? Estimate the stopping-power ratio data from Harder's approximation, see Chapter 9 (the mean energy at the polystyrene depth should be determined first).
*Answer: (a) 1.916 Gy; (b) 1.851 Gy.*

**Solution:**

(a) For a given charge collected in a mass of air, the absorbed dose in the chamber is given by

$$D_{air} = \frac{q}{m_{air}} (W/e)_{air}$$

where $q = q_0\, k_{TP}\, k_s$, with $q_0 = 64.2 \times 10^{-9}$ C and

$$k_{TP} = \frac{273.15 + T\,°C}{273.15 + T_{ref}\,°C} \frac{P_{ref}}{P} = \frac{273.15 + 25.2}{273.15 + 20.0} \frac{101.325}{102.0} = 1.0110$$

$$k_s = \frac{1}{0.973} = 1.0277$$

(note that the saturation correction factor is the inverse of the ion-collection efficiency). Therefore

$$q = 64.2 \times 10^{-9}\ \text{C} \times 1.0110 \times 1.0277 = 66.71 \times 10^{-9}\ \text{C}$$

For the chamber with $h = 0.2$ cm and $r = 1.25$ cm, its air volume is $v = 0.982$ cm³; the air mass is $m_{air} = 1.183 \times 10^{-6}$ kg for $\rho_{air} = 1.2045 \times 10^{-3}$ g cm⁻³. The absorbed dose in the chamber air results:

$$D_{air} = \frac{66.71 \times 10^{-9}\ \text{C}}{1.183 \times 10^{-6}\ \text{kg}} \times 33.97\ \text{J C}^{-1} = 1.916\ \text{Gy}$$

(b) The Bragg–Gray principle for this case can be written as

$$D_{polyst} = D_{air}\, s_{polyst,air}$$

where $s_{polyst,air}$ needs to be evaluated from Harder's approximation, Eq. (9.24) in the textbook, at the electron mean energy at a depth of $2\ \text{cm} \times \rho_{polyst} = 2\ \text{cm} \times 1.060$ g cm⁻³ = 2.12 g cm⁻²

$$\bar{E}_z = E_0 - z_{polyst}[S_{tot}(E_0)/\rho]_{polyst}$$

$$= 15\ \text{MeV} - 2.12\frac{\text{g}}{\text{cm}^2} \times 2.1962\ \text{MeV}\frac{\text{cm}^2}{\text{g}} = 10.34\ \text{MeV}$$

that is,

$$s_{polyst,air} \approx \frac{[S_{el}(\bar{E}_z)/\rho]_{polyst}}{[S_{el}(\bar{E}_z)/\rho]_{air}} = \frac{1.9193}{1.9865} = 0.9661$$

where the mass electronic stopping powers are obtained from the electronic Data Tables.

Therefore, the absorbed dose in polystyrene is

$$D_{polyst} = 1.916\ \text{Gy} \times 0.9661 = 1.851\ \text{Gy}$$

**5** A Farmer-type ionization chamber (cylindrical, 0.6 cm³) has been calibrated in a 6 MV pulsed photon beam at a standards laboratory and serves as reference instrument at a hospital. The calibration coefficient in terms of absorbed dose to water is $N_{D,w,Q_0} = 45.74$ mGy nC⁻¹, obtained at 10 cm depth, field size of 10 cm×10 cm, 20 °C, 101.325 kPa, and +/−250 V (that is, corrected for polarity). Another chamber similar in size but with different design is to be cross-calibrated at the hospital for routine use (field instrument). Measurements are made with both chambers in a 6 MV beam of identical quality to that at the standards laboratory (in terms of $TPR_{20,10}$ or $\%dd(10)_x$). Using the data provided below at two voltages and varying polarity, determine the calibration coefficient of the field instrument. The instruments' readings in nC are

$$M_{ref}(+250V) = 7.210, \quad M_{ref}(-250V) = 7.246, \quad M_{ref}(+80V) = 7.168$$

at 21.91 °C and 101.8 kPa, and

$$M_{field}(+250V) = 10.632, \quad M_{field}(-250V) = 10.690, \quad M_{field}(+80V) = 10.569$$

at 21.88 °C and 101.9 kPa. For the saturation correction factor, use the polynomial function for the two-voltage method,

$$k_s = a_0 + a_1 \, (M_{V_1}/M_{V_2}) + a_2 \, (M_{V_1}/M_{V_2})^2$$

with coefficients $a_i$ as a function of voltage ratio $V_1/V_2$ given in the table below.

| $V_1/V_2$ | 2 | 2.5 | 3 | 3.5 | 4 | 5 |
|---|---|---|---|---|---|---|
| $a_0$ | 2.337 | 1.474 | 1.198 | 1.080 | 1.022 | 0.975 |
| $a_1$ | −3.636 | −1.587 | −0.875 | −0.542 | −0.363 | −0.188 |
| $a_2$ | 2.299 | 1.114 | 0.677 | 0.463 | 0.341 | 0.214 |

*Answer: 38.19 mGy nC$^{-1}$.*

**Solution:**
Equation (15.6) in the textbook gives the general expression for the cross-calibration of a field instrument, which for the present case can be written as

$$N_{D,w,Q_0}^{field} = N_{D,w,Q_0}^{ref} \frac{M_{Q_0}^{ref}}{M_{Q_0}^{field}}$$

where the readings correspond to the calibration voltage and polarity used for the calibration of the reference chamber. The raw readings must be corrected for the relevant influence quantities, that is,

$$M^{corr} = M_{+250} \, k_{TP} \, k_{pol} \, k_s$$

where we have
− for temperature and pressure

$$k_{TP} = \frac{273.15 + T \,°C}{273.15 + T_{ref} \,°C} \frac{P_{ref}}{P}, \text{with } T_{ref} = 20 \,°C \text{ and } P_{ref} = 101.325 \text{ kPa}$$

– for polarity, readings at different polarities are required and

$$k_{pol} = \frac{|M_{V+}| + |M_{V-}|}{2|M|}$$

where $V$ is the applied voltage with a different polarity ($\pm 250$ V) and the denominator corresponds to the reading at the chosen user voltage (say $+250$ V).

– for recombination, readings at different voltages are required and

$$k_s = a_0 + a_1 (M_{V_1}/M_{V_2}) + a_2(M_{V_1}/M_{V_2})^2$$

where $V_1$ is the calibration voltage, $+250$ V, $V_2$ is the reduced voltage with the same polarity, $+80$ V, and the coefficients $a_i$ must be obtained by interpolation for the ratio $V_1/V_2 = 3.125$.

Summarizing the different data and calculations in Table 15.1 (all $M$-readings in nC) one gets

$$N_{D,w,6\ MV}^{field} = 45.74\ mGy/nC \times \frac{7.2624\ nC}{8.6988\ nC} = 38.19\ mGy/nC$$

at 20 °C, 101.325 kPa, and $+250$ V

**Table 15.1 Summary of the data for Exercise 5.**

| Chamber | Reference | | Field |
|---|---|---|---|
| $M_{+250}$ | 7.210 | | 8.632 |
| $T$ (°C) | 21.91 | | 21.88 |
| $P$ (kPa) | 101.8 | | 101.9 |
| $k_{TP}$ | 1.0018 | | 1.0007 |
| | | | |
| $M_{-250}$ | 7.246 | | 8.690 |
| $k_{pol}$ | 1.0025 | | 1.0034 |
| | | | |
| $V_{+250}/V_{+80}$ | | 3.125 | |
| $a_0$ | | 1.1685 | |
| $a_1$ | | −0.7918 | |
| $a_2$ | | 0.6235 | |
| $M_{+80}$ | 7.168 | | 8.569 |
| $M_{+250}/M_{+80}$ | 1.0059 | | 1.0074 |
| $k_s$ | 1.0029 | | 1.0036 |
| | | | |
| Therefore | | | |
| $M^{corr}$ | 7.2624 | | 8.6988 |

**6** A 10 MeV broad electron beam impinges on a water phantom where a plane-parallel ionization chamber is inserted. The chamber air cavity has

10 mm diameter and 2.5 mm height; it is irradiated along the cylinder axis with the front of its volume positioned at a depth of 2 cm. Scattering and attenuation effects in the chamber walls are assumed to be negligible. Estimate the dose to the air in the chamber if the electron fluence at 2 cm depth is $0.5 \times 10^{10}$ cm$^{-2}$.

*Answer: 2 Gy.*

**Solution:**

The mean energy of electrons reaching the chamber front surface at 2 cm depth is estimated using the total energy loss

$$\bar{E}_z = E_0 - z_w \, [S_{tot}(E_0)]_w$$

where, from the electronic Data Tables, $[S_{tot}(10\text{MeV})]_w = 2.1482$ MeV cm$^2$/g $= 2.1444$ MeV/cm for $\rho_w = 0.9982$ g/cm$^3$.

Thus

$$\bar{E}_z = 10 \text{ MeV} - 2 \text{ cm} \times 2.1444 \text{ MeV/cm} = 5.711 \text{ MeV}$$

The mean chord length $\bar{\ell}$ for the chamber air cavity with 0.5 cm radius and 0.25 cm height is

Area: $a = 2\pi r^2 + 2\pi r h = 2.3562$ cm

Volume: $v = \pi r^2 h = 0.1963$ cm$^3$

$$\bar{\ell} = \frac{4v}{a} = \frac{4 \times 0.1963}{2.3562} = 0.3333 \text{ cm}$$

The energy deposited in the cavity is estimated from the electronic stopping power in air evaluated at the mean energy of the electrons reaching the chamber front surface,

$$[S_{el}(5.711 \text{ MeV})]_{air} = 1.8599 \text{ MeV cm}^2 \text{ g}^{-1} = 2.2402 \times 10^{-3} \text{ MeV cm}^{-1}$$

for $\rho_{air} = 1.2045 \times 10^{-3}$ g cm$^{-3}$, that is,

$$E_{dep} = 2.2402 \times 10^{-3} \text{ MeV cm}^{-1} \times 0.3333 \text{ cm} = 7.4674 \times 10^{-4} \text{ MeV}.$$

As the air cavity has a mass

$$m_{air} = v \times \rho_{air} = 0.1963 \text{ cm}^3 \times 1.2045 \times 10^{-3} \text{ g cm}^{-3}$$
$$= 2.3650 \times 10^{-4} \text{ g} = 2.3650 \times 10^{-7} \text{ kg}$$

the absorbed dose per incident electron is

$$D_{air} = \frac{E_{dep}}{m_{air}} = \frac{7.4674 \times 10^{-4} \text{ MeV}}{2.3650 \times 10^{-7} \text{ kg}} \times 1.6022 \times 10^{-13} \frac{\text{J}}{\text{MeV}}$$
$$= 5.0588 \times 10^{-10} \text{ Gy}$$

As the electron fluence at 2 cm depth is $0.5 \times 10^{10}$ cm$^{-2}$, and the chamber front surface is $\pi r^2 = 0.7854$ cm$^2$, the absorbed dose for the fluence incident on the chamber results as follows:

$$D_{air} = 5.0588 \times 10^{-10} \frac{\text{Gy}}{\text{electron}} \times 0.5 \times 10^{10} \frac{\text{electron}}{\text{cm}^2} \times 0.7854 \text{ cm}^2$$
$$= 1.987 \text{ Gy} \approx 2 \text{ Gy}$$

**7** An NE-2571 Farmer-type ionization chamber has a $N_{D,w,^{60}Co}$ calibration coefficient of 47.54 mGy nC$^{-1}$. It is placed at a depth of 10 cm in a water phantom to calibrate an 8 MV (TPR$_{20,10}$ = 0.70) 10 cm $\times$ 10 cm photon beam, yielding a collected charge of $3.70 \times 10^{-8}$ C at $P$ = 102.3 kPa, $T$ = 22.5 °C for an ion-collection efficiency of 0.995 and negligible polarity effect. (a) Determine the absorbed dose at the calibration depth using the fit to experimental data for this chamber

$$k_{Q,^{60}Co} = \frac{193\ 077}{192\ 538 + \exp\ (11.1674 \times TPR_{20,10})}$$

(b) Assuming one would be interested in doing the measurements in a plastic phantom with composition (fraction by weight) H: 0.077418, C: 0.922582, and $\rho_{plastic}$ = 1.06 g cm$^{-3}$, at what depth should the chamber center be placed to simulate the 10 cm depth in water? (c) How would you determine the dose to water from measurements in the plastic phantom under the $N_{D,w}$ formalism?
*Answer: (a) 1.75 Gy, (b) 9.72 cm, (c) see Eq. (16.10) and related text.*

**Solution:**

(a) For the $N_{D,w}$ formalism, the absorbed dose to water is given by

$$D_{w,Q}(z_{ref,w}) = M_{w,Q} N_{D,w,Q_0} k_{Q,Q_0}$$

where $M_{w,Q} = q_{w,Q}\ k_{TP}\ k_{pol}\ k_s$, with

$$q_{w,Q} = 37.0 \times 10^{-9}\ C$$

$$k_{TP} = \frac{273.15 + 22.5}{273.15 + 20.0} \frac{101.325}{102.3} = 0.9989$$

$$k_{pol} = 1.0$$

$$k_s = 1/\text{collec eff} = 1/0.995 = 1.0050$$

Thus,

$$M_{w,Q} = 37.15 \times 10^{-9}\ C = 37.15\ nC$$

The beam quality factor related to a $Q_0 = {}^{60}Co$ for a TPR$_{20,10}$ = 0.70 is obtained from

$$k_{Q,^{60}Co} = \frac{193077}{192538 + e^{11.1674 \times TPR_{20,10}}} = 0.990$$

Therefore

$$D_{w,Q}(z = 10\ cm) = 37.15\ nC \times 47.54\ \frac{mGy}{nC} \times 0.990$$
$$= 1748.3\ mGy \approx 1.75\ Gy$$

**Table 15.2** Composition data for the plastic phantom of Exercise 7.

| Element | $Z_i$ | $w_i$ | $(Z/A)_i$ |
|---------|-------|----------|-----------|
| H | 1 | 0.077418 | 0.992162 |
| C | 6 | 0.922582 | 0.499542 |

(b) The depth in plastic equivalent to the reference depth in water is obtained by scaling according to the ratio of electron densities, see Eq. (16.11) in the textbook, that is,

$$z_{\text{plastic}} = \frac{\rho_w}{\rho_{\text{plastic}}} \times \frac{(Z/A)_w}{(Z/A)_{\text{plastic}}} \times z_w$$

where $(Z/A)_w = 0.555087$ (from the Data Tables) but $(Z/A)_{\text{plastic}}$ must be determined from the plastic composition and the $(Z/A)_i$ for the respective elements (see Table 15.2):

$$(Z/A)_{\text{plastic}} = \sum_i (w \times Z/A)_i$$

$$= 0.077418 \times 0.992162 + 0.922582 \times 0.499542 = 0.537680$$

Thus

$$z_{\text{plastic}} = \frac{0.9982}{1.06} \times \frac{0.555087}{0.537680} \times 10 \text{ cm} = 9.72 \text{ cm}$$

(c) Equation (16.10) in the textbook provides a modified expression for the absorbed dose to water from measurements in a plastic phantom having the form

$$D_{w,Q}(z_{\text{ref,w}}) = M_{\text{plastic},Q}(z_{\text{eq,plastic}}) \, N_{D,w,^{60}\text{Co}} \, k_{Q,^{60}\text{Co}} \, k_Q^{w,\text{plastic}}$$

where the conversion factor $k_Q^{w,\text{plastic}}$ must be determined theoretically (e.g., by a Monte Carlo simulation) or, preferably, by experiment. As an illustration, Figure 16.18 in the textbook provides approximate values for two plastic materials.

**8** Describe and provide the corresponding equation for the beam quality correction factor in the dosimetry of kilovoltage and megavoltage photon beams.
*Answer: $k_Q$ is defined as the ratio of air-kerma (kV) and absorbed dose to water (MV) calibration coefficients.*

**Solution:**
The beam quality correction factor corrects the calibration coefficient (not calibration factor, as it has units) for the difference in beam quality between the standard laboratory and the user's beam. It has the following forms:

(a) In kilovoltage dosimetry:

$$K_{air,Q} = M_Q \, N_{K,air,Q_0} \, k_{Q,Q_0}$$

where

$$k_{Q,Q_0} = \frac{N_{K,air,Q}}{N_{K,air,Q_0}}$$

The beam quality correction factor is defined as the ratio of air-kerma calibration coefficients at the user quality $Q$ and at the laboratory reference quality $Q_0$.

(b) In megavoltage dosimetry:

$$D_{w,Q} = M_Q \, N_{D,w,Q_0} \, k_{Q,Q_0}$$

where

$$k_{Q,Q_0} = \frac{N_{D,w,Q}}{N_{D,w,Q_0}}$$

The beam quality correction factor is defined as the ratio of absorbed dose to water calibration coefficients at the user quality $Q$ and at the laboratory reference quality $Q_0$ (usually $^{60}$Co $\gamma$ rays).

**9** Considering the general expression for the determination of absorbed dose to water in a user's radiotherapy beam, given by $D_{w,Q} = M_Q \, N_{S,Q_0} \, f_{Q,Q_0}^{D,S}$, explain the meaning of the three quantities in the right-hand side of the equation. Describe them in detail for the case of the formalism in terms of air kerma for high-energy photons and electrons, explaining the meaning of every quantity and factor in the equations.

**Solution:**
The general expression $D_{w,Q} = M_Q \, N_{S,Q_0} \, f_{Q,Q_0}^{D,S}$ states that the absorbed dose to water at the user's beam quality is obtained from the product of
(a) the instrument reading in the user's beam, $M_Q$, suitably corrected to the reference conditions for which $N_{S,Q_0}$ is valid (influence quantities)
(b) the calibration coefficient, $N_{S,Q_0}$, measured for the quantity $S$ (air kerma or absorbed dose) in the standards laboratory reference quality $Q_0$
(c) $f_{Q,Q_0}^{D,S}$ is the *overall* factor necessary to convert both from the calibration quantity $S$ to absorbed dose $D$ and from the calibration beam quality $Q_0$ to the user's beam quality $Q$.
In the case of the air-kerma formalism for high-energy photons and electrons, the expression can be written as

$$D_{w,Q} = M_Q \, N_{K,air,^{60}Co} \, f_{Q,^{60}Co}^{D,K_{air}}$$

where $N_{K,air,^{60}Co}$ is the air-kerma calibration coefficient at the reference beam quality of $^{60}$Co $\gamma$ rays.
Using the Bragg–Gray principle including ionization chamber perturbation correction factors, the equation can be written as (it is important to keep

track of factors and quantities for $^{60}$Co and for the user's beam quality $Q$)

$$D_{w,Q} = D_{air,Q} \, s_{w,air} \, p_{ch,Q} = M_Q \, N_{D,air} \, s_{w,air} \, p_{ch,Q}$$

$$= M_Q \left[ N_{K_{air}} (1 - \bar{g}_{air}) \prod_i k_i \right]_{^{60}Co} \left[ s_{w,air} \prod_j p_j \right]_Q$$

and grouping

$$D_{w,Q} = M_Q \, N_{K,air,^{60}Co} \left[ (1 - \bar{g}_{air}) \prod_i k_i \right]_{^{60}Co} \left[ s_{w,air} \prod_j p_j \right]_Q$$

that is,

$$f_{Q,^{60}Co}^{D,K_{air}} = \left[ (1 - \bar{g}_{air}) \prod_i k_i \right]_{^{60}Co} \left[ s_{w,air} \prod_j p_j \right]_Q$$

The meaning of the different quantities and factors in the expression above is:

(a) $N_{D,air}$ is the cavity air calibration coefficient, used to get the dose to the air of the cavity when multiplied by $M_Q$. It is defined at the $^{60}$Co calibration quality as

$$N_{D,air,^{60}Co} = \left[ N_{K_{air}} (1 - \bar{g}_{air}) \prod_i k_i \right]_{^{60}Co} = N_{K_{air}} (1 - \bar{g}_{air}) \, k_{att} \, k_m \, k_{cel}$$

where (all at $^{60}$Co γ rays)

$\bar{g}_{air}$    is the fraction of the initial kinetic energy of the secondary electrons generated by $^{60}$Co γ rays, which is converted into bremsstrahlung in air

$k_m$    corrects for the non-air equivalence of the chamber wall and build-up cap used during the air-kerma calibration

$k_{att}$    corrects for attenuation and scatter in the chamber wall and build-up cap

$k_{cel}$    corrects for the non-air equivalence of the chamber central electrode

It should be emphasized that $N_{D,air}$ is constant for a given ionization chamber, that is, it is an indirect measure of its volume provided that $W_{air}/e$ does not vary with energy within the megavoltage range (usual approximation for photon and electron beams). This means that $N_{D,air}$ can be used at any other energy in this range.

(b) $s_{w,air}$ is the restricted stopping-power ratio (i.e., the Spencer–Attix) water/air averaged over the electron spectra (produced either by primary electrons or by photons) at the depth of measurement in the user's beam of quality $Q$.

(c) $p_{ch,Q}$ is an overall chamber perturbation correction factor at the quality $Q$, which accounts for the perturbation of the particle fluence caused by the presence of the ionization chamber in the medium (water). It is given by the product of several correction factors, each accounting for a different perturbation effect, assumed to be independent of each other:

$$p_{\text{ch},Q} = \left[ \prod_j p_j \right]_Q = [p_{\text{dis}}\, p_{\text{wall}}\, p_{\text{cel}}\, p_{\text{cav}}]_Q$$

where

$p_{\text{dis}}$  accounts for the effect of replacing a volume of water with the chamber air cavity in cylindrical chambers

$p_{\text{wall}}$  accounts for the non-water equivalence of the chamber wall and any waterproofing material

$p_{\text{cel}}$  accounts for the non-air equivalence of the central electrode during in-phantom measurements

$p_{\text{cav}}$  accounts for the effects of the air cavity on the in-scattering of electrons making the electron fluence different from that in water in the absence of the cavity

# 16

# Dosimetry of Small and Composite Radiotherapy Photon Beams

1  The condition of charged particle non-equilibrium in small photon fields
   (a) only manifests itself in dosimetry measurements using an air-filled chamber in small fields
   (b) is a condition that reduces absorbed dose on the central axis of small fields
   (c) affects the stopping-power ratio water to air in small fields
   (d) is present, regardless of whether a detector is used in the field
   (e) makes beam output measurements with air-filled ionization chambers in small fields possible
   *Answer: (1d): is present, regardless of whether a detector is used in the field.*

2  The stopping-power ratio water-to-air in a 5 mm × 5 mm field differs from the value in a 10 cm × 10 cm field by
   (a) < 0.5%
   (b) 1%
   (c) 2%
   (d) 5%
   (e) >5%
   *Answer: (2a) < 0.5%.*

3  Unshielded silicon diode detectors show field size-dependent correction factors due to two competing effects:
   (a) Intrinsic energy dependence of Si and volume averaging
   (b) Intrinsic energy dependence of Si and perturbation effects
   (c) Polarity effect and recombination
   (d) Polarity effect and electrometer calibration
   (e) Recombination effect and diode doping
   *Answer: (3b): Intrinsic energy dependence of Si and perturbation effects.*

4  Indicate the two largest contributors to the correction factors and their uncertainties for commercial air-filled ionization chambers in small photon fields.
   (a) The stopping-power ratio and the central electrode effect
   (b) The stopping-power ratio and the chamber wall effect
   (c) The density perturbation effect and the volume averaging effect

*Fundamentals of Ionizing Radiation Dosimetry: Solutions to Exercises,* First Edition.
Pedro Andreo, David T. Burns, Alan E. Nahum, and Jan Seuntjens.
© 2017 Wiley-VCH Verlag GmbH & Co. KGaA. Published 2017 by Wiley-VCH Verlag GmbH & Co. KGaA.

    (d) The stopping-power ratio and the volume averaging effect

    (e) The ionization chamber wall effect and the stem effect

*Answer: (4c) The density perturbation effect and the volume averaging effect.*

**5** Beam quality specification and the explicit measurement of the beam quality specifier in small-field dosimetry are required in the following instance:

    (a) To specify the field correction factor applied to the measured output ratios to obtain output factors

    (b) To specify small field output factors

    (c) To specify the $k_{Q_{msr},Q}^{f_{msr},f_{ref}}$ beam quality correction factor in the *msr* field

    (d) To ensure the beam is of adequate quality

    (e) To specify the absorbed dose calibration coefficient for a small field

*Answer: (5c): To specify the $k_{Q_{msr},Q}^{f_{msr},f_{ref}}$ beam quality correction factor in the msr field.*

**6** Determination of the equivalent-square field size for a machine-specific reference field involves:

    (a) Using the geometric mean of width and length of a not too elongated field

    (b) Considering the field that produces an equal amount of scatter on the central axis at the measurement depth as the circular or rectangular field

    (c) Considering manufacturer-imposed restriction on the choice of field size

    (d) Equating the length of the largest side of the rectangular field size with the diameter of a circular field

*Answer: (6b): One must consider the field that produces an equal amount of scatter on the central axis at the measurement depth as the circular or rectangular field.*

**7** To determine the absorbed dose to water from a measurement made with a detector in a plastic phantom:

    (a) A phantom dose-conversion factor of unity is acceptable

    (b) The depth must be scaled by the ratio of the electron densities of the plastic versus water

    (c) The field size must be scaled by the ratio of the electron densities of the plastic versus water

    (d) The SSD must be scaled by the ratio of the electron densities of the plastic versus water

    (e) All of (b)-(d) must be done or accounted for

*Answer: (7e): All of (b)-(d) must be done or accounted for.*

# 17

# Reference Dosimetry for Diagnostic and Interventional Radiology

Note that Exercises 1–5 of Chapter 7, related to the spectra and beam quality of kV x-ray beams, are applicable to this chapter.

1  A monoenergetic photon beam has a mass attenuation coefficient of $\mu/\rho = 1.3086$ cm$^2$ g$^{-1}$ in rhodium. What is the average distance that a photon in this beam will travel in rhodium before interacting?
*Answer: 0.6 mm.*

**Solution:**
The question corresponds to the definition of mean free path, equal to the inverse of the linear attenuation coefficient.
Considering that the density of rhodium is $\rho = 12.41$ g cm$^{-3}$, the linear attenuation coefficient is

$$\mu = \mu/\rho \times \rho = 1.3086\frac{\text{cm}^2}{\text{g}} \times 12.41\frac{\text{g}}{\text{cm}^3} = 16.24 \text{ cm}^{-1}$$

from which

$$\text{MFP} = 1/\mu = 0.06 \text{ cm} = 0.6 \text{ mm}$$

2  A narrow monoenergetic photon beam has a first half-value layer of 4 mm in aluminium. By how much will the kerma decrease after the beam has traversed 3 cm of water?
*Answer: 43% approximately.*

**Solution:**
Recall that the relation between HVL and the mass attenuation coefficient is obtained from the exponential attenuation expression

$$K(t) = K(0) \, e^{-\mu/\rho \times t}$$

with the absorbing material thickness $t$ in g cm$^{-2}$. Making $K(t) = K(0)/2$, as inferred from the definition of HVL, gives

$$\frac{1}{2} = e^{-\mu/\rho \times \text{HVL}}$$

*Fundamentals of Ionizing Radiation Dosimetry: Solutions to Exercises,* First Edition.
Pedro Andreo, David T. Burns, Alan E. Nahum, and Jan Seuntjens.
© 2017 Wiley-VCH Verlag GmbH & Co. KGaA. Published 2017 by Wiley-VCH Verlag GmbH & Co. KGaA.

from which, taking logarithms on both sides, results in

$$\mu/\rho = \frac{\ln 2}{HVL}$$

with HVL expressed in g cm$^{-2}$. Hence, considering that the density of aluminium is $\rho = 2.699$ g cm$^{-3}$, we obtain the result

$$(\mu/\rho)_{Al} = \frac{\ln 2}{HVL\,\rho} = \frac{\ln 2}{0.4\text{ cm} \times 2.699\text{ g/cm}^3} = 0.6420\text{ cm}^2/\text{g}$$

From the Data Tables, this aluminium mass attenuation coefficient corresponds approximately to a photon energy $k = 37.8$ keV, which in water corresponds to $(\mu/\rho)_w = 0.2844$ cm$^2$ g$^{-1}$.
Therefore, as $\rho_w = 0.9982$ g cm$^{-3}$

$$K(3\text{ cm})_w = K(0)_w\, e^{-(\mu/\rho)_w t_w}$$

$$\frac{K(3\text{ cm})_w}{K(0)_w} = \exp[-0.2844\text{ cm}^2/\text{g} \times 3\text{ cm} \times 0.9982\text{ g/cm}^3] = 0.4267$$

that is, 42.67%.

3   An x-ray unit operating at a potential of 150 kV delivers an exposure of 100 mA during 2 s. Assuming that all the incident electron energy is absorbed as electrical energy in the target, estimate the amount of energy deposited (J) and the subsequent rise in temperature (°C) in a tungsten target of 12 cm$^2$ area and 3 mm thickness. The specific heat capacity of tungsten (at 20 °C) is 134 J kg$^{-1}$ K$^{-1}$, and its density is 19.30 g cm$^{-3}$.
*Answer: $3 \times 10^4$ J; approximately 3200 °C.*

**Solution:**
The energy deposited in the target can, in a first approximation, be estimated in terms of the absorbed electrical energy $E_{el} = q\,V = i\,t\,V$, that is,

$$E_{el} = i\,t\,V = 0.1\text{ A} \times 2\text{ s} \times 10^3\text{ V} = 3 \times 10^4\text{ J}$$

The rise in temperature is given by (see, e.g., Eq. (11.18) in the textbook)

$$\Delta T = \frac{E_{el}}{\text{mass} \times c_w}$$

For the present exercise, we have
$t = 0.3$ cm, area $= 12$ cm$^2 \Rightarrow$ vol $= 3.6$ cm$^3 \Rightarrow$ mass $= \rho$ vol $= 69.48$ g
Hence

$$\Delta T = \frac{3 \times 10^4\text{ J}}{69.48\text{ g} \times 10^{-3}\,\frac{\text{kg}}{\text{g}} \times 134\,\frac{\text{J}}{\text{kg°C}}} = 3222.2\text{ °C}$$

Note that even if in reality there are heat losses due to conduction, and some energy is taken away by photons, this temperature is close to the melting point of tungsten (about 3400 °C), pointing at the need for rotating anodes. Alternatively, we can determine the number of 150 keV electrons that correspond to 100 mA and 2 s, see next exercise. As the target is rather thick (compared with the $R_{CSDA}$ of 150 keV electrons, 0.003 cm), most of the energy of these electrons will be absorbed, but still a substantial amount of photons will emerge from it.

**4** For the conditions used in the previous exercise (150 kV, 100 mA, 2 s) estimate the amount of bremsstrahlung x-ray energy (J) generated by 150 keV electrons in (a) a tungsten target 0.01 mm thick and (b) a thick tungsten target (total absorption of electrons). (c) Estimate the total electron energy deposited in the previous exercise assuming that all of it is absorbed in the target.
*Answer: (a) 177.6 J; (b) 441 J; (c) 3 × 10⁴ J.*

**Solution:**
We determine first the number of 150 keV electrons that correspond to the conditions used. From the Data Tables, the electron charge is $e = 1.6022 \times 10^{-19}$ C, which means that there are $n_{e,C}$ electrons per coulomb:

$$n_{e,C} = 6.2415 \times 10^{18} \text{ elect/C}$$

The current is $i = 0.1$ A $= 0.1$ C $s^{-1}$; the number of 150 keV electrons in 2 s therefore is

$$n_e = n_{e,C}\, i\, t = 6.2415 \times 10^{18} \text{ elect/C} \times 0.1 \text{ C } s^{-1} \times 2 \text{ s}$$
$$= 1.2483 \times 10^{18} \text{ electrons}$$

From the electronic Data Tables, the radiative stopping power of 150 keV electrons in tungsten is $S_{rad}/\rho = 0.0460$ MeV cm$^2$ g$^{-1}$, the radiation yield is $Y = 0.0147$, and the CSDA range is $R_{CSDA} = 0.0567$ g cm$^{-2} = 0.0034$ cm. Hence,

(a) A tungsten target thickness of 0.01 mm is approximately 1/3 of the electron CSDA range, which means that electrons will traverse the target losing radiation energy by bremsstrahlung in it (in addition to the electronic energy losses). For the $n_e$ 150 keV electrons, this energy loss will be given by the radiative stopping power within the target thickness, that is,

$$R = n_e\, S_{rad}/\rho\, t\rho$$
$$= 1.2483 \times 10^{18} \text{ elect} \times 0.046\, \frac{\text{MeVcm}^2\text{g}^{-1}}{\text{elect}} \times 0.001\text{cm} \times 19.30\frac{\text{g}}{\text{cm}^3}$$
$$\times 1.6022 \times 10^{-13}\frac{\text{J}}{\text{MeV}}$$
$$= 177.56 \text{ J}$$

(b) In a thick target ($t \geq R_{CSDA}$), all the electron energy will be absorbed, and the relevant quantity (fraction of energy going into bremsstrahlung) is the radiation yield, that is,

$$R = n_e\, Y\, E$$
$$= 1.2483 \times 10^{18} \text{ elect} \times 0.0147 \times 0.150\, \frac{\text{MeV}}{\text{elect}} \times 1.6022 \times 10^{-13}\, \frac{\text{J}}{\text{MeV}}$$
$$= 441.01 \text{ J}$$

(c) The target thickness is approximately $3 \times R_{CSDA}$. Assuming that all the electron energy is absorbed in it,

$$\Delta E = n_e \, E$$

$$= 1.2483 \times 10^{18} \text{ elect} \times 0.150 \, \frac{\text{MeV}}{\text{elect}} \times 1.6022 \times 10^{-13} \, \frac{\text{J}}{\text{MeV}}$$

$$= 3 \times 10^4 \text{ J}$$

as in the previous exercise.

5  Determine (a) the reference air kerma free-in-air for measurements made in air at a focus–detector distance of 60 cm with a reference dosimeter having a calibration coefficient for an RQR-5 beam (70 kV and $HVL_1 = 2.58$ mm Al) of $3 \times 10^5$ Gy $C^{-1}$ at 20°C, and 101.325 kPa. The detector readings in nC are 1.006, 1.000, 0.994, 0.998, and 0.993, all at 22.1°C and 101.1 kPa. (b) Calculate the incident air kerma on a phantom 25 cm thick, placed on a table at a distance to the focus of 100 cm.
*Answer: (a) 0.30 mGy; (b) 0.19 mGy.*

**Solution:**

(a) The reference air-kerma free-in-air is given by

$$K_{air,ref} = M_{air} \, k_{TP} \, N_{K,air}$$

where
- $M_{air}$ is the average of the detector readings, equal to 0.998 nC
- $k_{TP} = \dfrac{273.15 + T°C}{273.15 + T°_{ref}C} \dfrac{P_{ref}}{P} = \dfrac{273.15 + 22.1}{273.15 + 20} \dfrac{101.325}{101.1} = 1.0094$, and
- $N_{K,air}$ is the detector calibration coefficient, which can be written as $N_{K,air} = 3 \times 10^5$ Gy $C^{-1} = 0.3$ mGy $nC^{-1}$

These yield

$$K_{air,ref} = 0.998 \text{ nC} \times 1.0094 \times 0.3 \, \frac{\text{mGy}}{\text{nC}} = 0.302 \text{ mGy}$$

(b) The reference distance is $d_{ref} = 60$ cm, and the distance to the phantom surface is $d = 100 - 25 = 75$ cm. Using the inverse square law yields

$$K_{air,i} = K_{air,ref} \left( \frac{d_{ref}}{d} \right)^2 = 0.302 \text{ mGy} \left( \frac{60 \text{ cm}}{75 \text{ cm}} \right)^2 = 0.193 \text{ mGy}$$

6  Use the data from the previous exercise to cross-calibrate in terms of air kerma–area product ($P_{KA}$) a KAP-meter whose readings at the reference distance and under the same $P$ and $T$ conditions as above are 0.965 nC, 0.968 nC, 0.975 nC, 0.970 nC, and 0.967 nC. The beam area at the measurement reference position is 6 cm $\times$ 6 cm. Give $N_{P_{KA}}$ in $\mu$Gy m$^2$ nC$^{-1}$.
*Answer: 1.113 $\mu$Gy m$^2$ nC$^{-1}$.*

**Solution:**

For a cross-calibration in terms of air kerma–area product $P_{KA}$, we can write the expr ession

$$N_{P_{KA,Q}} = \frac{P_{KA,Q}^{\text{ref}}}{M_Q^{\text{KAP}} k_{TP}} \equiv \frac{K_{\text{air,ref}} A}{M_Q^{\text{KAP}} k_{TP}}$$

where

- $K_{\text{air,ref}}$ is the reference air kerma free-in-air determined in the previous exercise,
- $A$ is the beam area at the measurement position, their product yielding the air kerma–area product $P_{KA}$, and
- $M_Q^{\text{KAP}} k_{TP}$ is the average KAP-meter reading corrected for influence quantities ($T$ and $P$).

For simplicity, we write all the quantities in the desired final units, that is, $K_{\text{air,ref}} = 302.25$ µGy and $A = 0.0036$ m$^2$, yielding $P_{KA} = 1.088$ µGy m$^2$, and $M_Q^{\text{KAP}} = 0.969$ nC with $k_{TP} = 1.009$ as in the previous exercise. Substituting these values in the expression above, we get

$$N_{P_{KA,Q}} = \frac{302.25 \text{ µGy} \times 0.0036 \text{ m}^2}{0.969 \text{ nC} \times 1.009} = 1.1126 \text{ µGy m}^2 \text{ nC}^{-1}$$

It should be noted that $N_{P_{KA,Q}}$ has been determined in a condition where the beam *incides on* the KAP detector. Often in practice, when the position of the KAP meter is at the exit of the generator head, $N_{P_{KA,Q}}$ should be determined for the *transmitted* beam, that is, $K_{\text{air,ref}}$ should be decreased by a factor taking into account the beam attenuation in the KAP-meter (usually determined experimentally). Some laboratories providing $N_{P_{KA,Q}}$ calibrations of KAP-meters express them in terms of incident and/or transmitted radiation.

It is also common to insert the value of $N_{P_{KA,Q}}$ into the software attached to the KAP detector, so that the user gets a KAP-meter reading directly in terms of µGy m$^2$ (or a multiple of it) instead of the conventional reading in nC.

7 For a 90 kV, HVL = 6 mm Al x-ray beam, a reference detector has a beam quality factor $k_{Q,Q_0} = 0.968$ relative to a reference quality for which $N_{K,Q_0} = 5 \times 10^5$ Gy C$^{-1}$ at 20 °C and 101.325 kPa. Measurements made free-in-air in a 10 cm × 10 cm field at 100 cm focus-to-detector distance, with $P = 103$ kPa and $T = 23.2$ °C are 100.4 nC s$^{-1}$, 100.0 nC s$^{-1}$, 99.59 nC s$^{-1}$, 99.88 nC s$^{-1}$, and 99.51 nC s$^{-1}$. Determine the entrance-surface absorbed dose rate in (a) a water phantom and (b) a PMMA phantom, both having a thickness of 8 cm. Approximate the necessary data from the figures in the chapter of the textbook.

*Answer: (a) 70.43 mGy s$^{-1}$; (b) 75.66 mGy s$^{-1}$.*

**Solution:**

(a) The entrance-surface dose rate on a water phantom (or water kerma in water) is given by the general expression

$$\dot{D}_{w,Q} = (\dot{K}_{w,Q})_w = \dot{M}_{air,Q}\, k_{TP}\, N_{K,\,air,Q}\, k_{Q,Q_0}\, B_{air}(Q)\, [\mu_{en}(Q)/\rho]_{w,air}$$

where

- $\dot{M}_{air}$ is the average of the detector readings, equal to 99.87 nC s$^{-1}$,
- $k_{TP} = \dfrac{273.15+T\ ^\circ C}{273.15+T_{ref}\ ^\circ C}\dfrac{P_{ref}}{P} = \dfrac{273.15+23.2}{273.15+20}\dfrac{101.325}{103} = 0.9945$
- $N_{K,air}$ is the detector calibration coefficient, $N_{K,air} = 5 \times 10^5$ Gy C$^{-1}$ = 0.5 mGy nC$^{-1}$
- $k_{Q,Q_0} = 0.968$ is the beam quality factor
- $B_{air}(Q) = 1.42$, from Figure 17.9 in the textbook, is the backscatter factor, and
- $[\mu_{en}(Q)/\rho]_{w,air} = 1.04$, from Figure 17.10 in the textbook, is the ratio of mass energy-absorption coefficients water-to-air.

These yield $\dot{D}_{w,Q} = 70.99$ mGy s$^{-1}$, which can be compared with the air-kerma rate free-in-air

$$(\dot{K}_{air,Q})_{air} = \dot{M}_{air,Q}\, k_{TP}\, N_{K,\,air,Q}\, k_Q = 48.07 \text{ mGy s}^{-1}$$

illustrating the important contribution of the backscatter factor $B_{air}(Q)$ to the entrance-surface dose and the much smaller contribution of the $\mu_{en}$-ratio.

The entrance-surface dose rate corresponds to a phantom thickness of 15 cm, that is,

$$\dot{D}_{w,Q}(15 \text{ cm}) = 70.99 \text{ mGy s}^{-1}$$

and to account for the backscatter reduction in a phantom 8 cm thick, we need to include the factor $f_{t,Q}$, which can be obtained from Figure 17.13 in the textbook by interpolation for a 10 cm × 10 cm field size. Our estimate gives a value of 0.992. Hence

$$\dot{D}_{w,Q}(8 \text{ cm}) = \dot{D}_{w,Q}(15 \text{ cm})\, f_{t,Q} = 70.99 \text{ mGy/s} \times 0.992 = 70.43 \text{ mGy/s}$$

(b) For the PMMA phantom, as a first step we would need to use $B_{air}(Q)_{PMMA}$ and $[\mu_{en}/\rho]_{PMMA,air}$ in the general equation. Instead of the backscatter factor, we use the phantom material factor $f_{m,Q}$, which for PMMA is given in Figure 17.12 of the textbook; for the current conditions, we get a value of $f_{PMMA,Q} = 1.07$. Due to the small variation in the $\mu_{en}$-ratio, we can assume that $[\mu_{en}/\rho]_{PMMA,air} \approx [\mu_{en}/\rho]_{w,air}$. Hence

$$\dot{D}_{PMMA,Q}(15 \text{ cm}) = \dot{D}_{w,Q}(15 \text{ cm})\, f_{PMMA,Q}$$
$$= 70.99 \text{ mGy s}^{-1} \times 1.07 = 75.96 \text{ mGy s}^{-1}$$

To account for the influence of an 8 cm PMMA phantom thickness, we can use the $f_{t,Q}$ water data for a water equivalent thickness of

$$t_w = t_{PMMA}\frac{\rho_{PMMA}}{\rho_w}\frac{(Z/A)_{PMMA}}{(Z/A)_w} = 8 \times \frac{1.19}{0.9982} \times \frac{0.539369}{0.555087} \approx 9.3 \text{ cm}$$

where the densities and $Z/A$ values for PMMA and for water are taken from the Data Tables. Using this thickness in Figure 17.13 of the textbook and interpolating for a 10 cm × 10 cm field, we estimate $f_{t,Q} = 0.996$. Hence

$$\dot{D}_{\text{PMMA},Q}(8 \text{ cm}) = \dot{D}_{\text{PMMA},Q}(15 \text{ cm}) f_{t,Q}$$
$$= 75.96 \text{ mGy s}^{-1} \times 0.996 = 75.66 \text{ mGy s}^{-1}$$

**8**  Measurements free-in-air in a CT scanner operating at 120 kV made with a KLP meter (pencil-type ionization chamber) provide an average reading of 77.7 nC, at $P = 100.8$ kPa and $T = 22.3 \,°C$, for 25 tomographic slices 7 mm thick. The detector calibration coefficient at a reference beam quality is $N_{P_{\text{KL},Q_0}} = 13.6$ mGy cm nC$^{-1}$ at 20 °C and 101.325 kPa, and, for the quality used, $k_{Q,Q_0} = 1.005$. (a) Determine the air-kerma length product $P_{\text{KL}}$ (often referred to as dose length product, DLP) and the CT air-kerma index $C_K$ (or CT dose index, CTDI). (b) Calculate the effective dose for a CT head examination if the $P_{\text{KL}}$ organ dose conversion coefficient is $c_{\text{head},P_{\text{KL}}} = 0.0023$ mSv mGy$^{-1}$cm$^{-1}$.

*Answer: (a) $P_{KL} = 1076$ mGy cm; (b) $C_K = 61.5$ mGy; (c) $D_{eff} = 2.5$ mSv.*

**Solution:**

(a)  The air-kerma length product $P_{\text{KL}}$ is determined by

$$P_{\text{KL},Q} = M_Q \, k_{\text{TP}} \, N_{P_{\text{KL},Q_0}} \, k_{Q,Q_0}$$

where
- $M_Q = 77.7$ nC
- $k_{\text{TP}} = \dfrac{273.15 + T\,°C}{273.15 + T_{\text{ref}}\,°C} \dfrac{P_{\text{ref}}}{P} = \dfrac{273.15 + 22.3}{273.15 + 20} \dfrac{101.325}{100.8} = 1.013$
- $N_{P_{\text{KL},Q_0}} = 13.6$ mGy cm nC$^{-1}$, and
- $k_{Q,Q_0} = 1.005$

These yield

$$P_{\text{KL},Q} \equiv \text{DLP} = 77.7 \text{ nC} \times 1.013 \times 13.6 \text{ mGy cm nC}^{-1} \times 1.005$$
$$= 1075.9 \text{ mGy cm}$$

(b)  The CT air-kerma index $C_K$ is given by

$$C_{K,Q} = \frac{P_{\text{KL},Q}}{N_i \, T_i}$$

where $N_i$ is the number of slices (25) and $T_i$ their thickness (0.7 cm). Hence

$$C_{K,Q} \equiv \text{CTDI} = \frac{1075.9 \text{ mGy cm}}{25 \times 0.7 \text{ cm}} = 61.48 \text{ mGy}$$

Note that, as by definition $P_{\text{KL}} = K_{\text{air}} \, t$ (with $t = N_i \, T_i$), the CT air-kerma index $C_{K,Q}$ coincides with $K_{\text{air}} = P_{\text{KL}}/t$ if the air kerma is assumed to be constant over the total scan length.

(c) The effective dose is in this case (see indices of $c_{organ,q}$) related to $P_{KL}$, that is,

$$D_{eff}(head) = P_{KL,Q} \, c_{head,P_{KL}}$$
$$= 1075.9 \text{ mGy cm} \times 0.0023 \text{ mSv mGy}^{-1} \text{ cm}^{-1}$$
$$= 2.47 \text{ mSv}$$

**9** The shutter transit time of a kV x-ray unit can be determined comparing a single exposure in a nominal time of t seconds with $n$ exposures of $t/n$ seconds. (a) Derive an expression for the estimation of the correction required and (b) apply it to the case of one exposure of 10 s yielding $K_1 = 50$ mGy and five exposures of 2 s yielding 11.95 mGy, 11.91 mGy, 11.86 mGy, 11.89 mGy, and 11.85 mGy.
*Answer: (a) $\Delta t = t \, (K_1 - K_n)/(K_n - n \, K_1)$; (b) 0.5 s.*

**Solution:**

(a) In the single-exposure case, the true air-kerma rate would be given by the measured air kerma divided by the measurement time, that is,

$$\dot{K} = \frac{K_1}{t + \Delta t}$$

where $\Delta t$ is the shutter transit time, which can be referred to as the *shutter error*.

In the case of multiple exposures, each measurement will include the $\Delta t$ error and the true air-kerma rate would be

$$\dot{K} = \frac{K_n}{t + n \, \Delta t}$$

where $K_n$ is the cumulative air kerma in the $n$ exposures of $t/n$ seconds, $K_n = \sum_{i=1}^{n} K_i$. Solving the equation

$$\frac{K_1}{t + \Delta t} = \frac{K_n}{t + n \, \Delta t}$$

for $\Delta t$, results in

$$\Delta t = t \, \frac{K_1 - K_n}{K_n - n \, K_1}$$

(b) For the present exercise, we have
- $t = 10$ s
- $K_1 = 50$ mGy
- $n = 5$
- $K_n = 11.95 + 11.91 + 11.86 + 11.89 + 11.85 = 59.58$ mGy

Hence

$$\Delta t = 10 \, \frac{50 - 59.58}{59.58 - 5 \times 50} = 0.503 \approx 0.5 \text{ s}$$

# 18

# Absorbed-Dose Determination for Radionuclides

1 Determine (a) the specific (or mass) activity of a $^{60}$Co source, comparing the resulting value with that given in Table 18.6 of the textbook and (b) the mass of a 100 MBq $^{137}$Cs source.
*Answer: (a) 42.61 TBq g$^{-1}$; (b) 30.17 μg.*

**Solution:**

(a) The specific activity $\mathcal{A}_m$ is defined by Eq. (18.13), that is,

$$\mathcal{A}_m = \frac{\ln 2}{T_{1/2}M} N_A$$

where
$T_{1/2} = 5.271$ y is the half-life of $^{60}$Co, see Table 18.2 of the textbook
$M = 58.933$ g mol$^{-1}$ is the molar atomic mass, see value of $A$ in the Data Tables
$N_A = 6.02214 \times 10^{23}$ mol$^{-1}$ is the Avogadro constant
$1$ y $\approx 31\ 536\ 000$ s
Therefore

$$\mathcal{A}_m = \frac{0.69315}{5.271 \text{ y} \times 58.933 \text{ g mol}^{-1}} \times 6.0221 \times 10^{23} \frac{1}{\text{mol}} \times \frac{1 \text{ y}}{31\ 536\ 000 \text{ s}}$$
$$= 42.61 \times 10^{12} \text{ Bq g}^{-1} = 42.61 \text{ TBq g}^{-1}$$

Table 18.6 in the textbook gives $0.042 \times 10^6$ GBq g$^{-1}$

(b) As mentioned earlier, but for $^{137}$Cs
$T_{1/2} = 30.05$ y
$M = 132.905$ g mol$^{-1}$
The specific activity is

$$\mathcal{A}_m = \frac{0.69315}{30.05 \text{ y} \times 132.905 \text{ g mol}^{-1}} \times 6.0221 \times 10^{23} \frac{1}{\text{mol}} \times \frac{1 \text{ y}}{31\ 536\ 000 \text{ s}}$$
$$= 3.314 \times 10^{12} \text{ Bq g}^{-1} = 3.314 \times 10^6 \text{ MBq g}^{-1}$$

and

$$\text{mass} = \frac{100 \text{ MBq}}{\mathcal{A}_m} = 3.017 \times 10^{-5} \text{ g} = 30.17 \text{ μg}$$

*Fundamentals of Ionizing Radiation Dosimetry: Solutions to Exercises*, First Edition.
Pedro Andreo, David T. Burns, Alan E. Nahum, and Jan Seuntjens.
© 2017 Wiley-VCH Verlag GmbH & Co. KGaA. Published 2017 by Wiley-VCH Verlag GmbH & Co. KGaA.

**2** Determine the air-kerma rate constants, $\Gamma_\delta$ for cut-off energies of 5 keV and 20 keV, and the corresponding $\Gamma_X$ values, for $^{59}$Fe, including the relevant data for $\bar{g}_{air}$. Compare the value obtained for $\Gamma_{20}$ with that given in Table 18.2 in the textbook. Use the disintegration scheme provided by the BIPM-LNHB database at http://www.nucleide.org/DDEP_WG/DDEPdata.htm.
*Answer:* $\Gamma_5 = 147.40$ *and* $\Gamma_{10} = 147.32\, m^2\, \mu Gy\, GBq^{-1}\, h^{-1}$; $\Gamma_X = 0.6221\, R\, m^2\, Ci^{-1}\, h^{-1}$.

**Solution:**
The air-kerma rate constant is given by Eq. (18.20) of the textbook in the form

$$\Gamma_\delta = 4590 \sum_i k_i\, n_i \frac{[\mu_{en}(k_i)/\rho]_{air}}{1 - \bar{g}_{air}} \quad m^2\, \mu Gy\, GBq^{-1}\, h^{-1}$$

where $k_i$ is expressed in MeV and $(\mu_{en}/\rho)_{air}$ in cm$^2$ g$^{-1}$.
We present the data using the same format as in Table 18.1 of the textbook, taking the disintegration scheme from the internet site and $(\mu_{en}/\rho)_{air}$ and $\bar{g}_{air}$ from the electronic Data Tables. For $\delta = 5$ keV, we have

| Energy, $k_i$ (keV) | Intensity (%) $100\, n_i$ | Type | $[\mu_{en}/\rho(k_i)]_{air}$ (cm$^2$ g$^{-1}$) | $1 - \bar{g}_{air}$ | $k_i\, n_i \frac{[\mu_{en}(k_i)/\rho]_{air}}{1-\bar{g}_{air}}$ |
|---|---|---|---|---|---|
| 6.9154 | 0.00596 | XKa2 | 1.438E+01 | 1.0000 | 5.926E−06 |
| 6.9304 | 0.0117 | XKa1 | 1.428E+01 | 1.0000 | 1.158E−05 |
| 7.6495 | 0.00243 | XKb1 | 1.056E+01 | 1.0000 | 1.963E−06 |
| 142.651 | 0.972 | g | 2.462E−02 | 0.9999 | 3.414E−05 |
| 192.349 | 2.918 | g | 2.643E−02 | 0.9997 | 1.484E−04 |
| 334.8 | 0.264 | g | 2.903E−02 | 0.9993 | 2.568E−05 |
| 382.0 | 0.0215 | g | 2.938E−02 | 0.9992 | 2.415E−06 |
| 1099.245 | 56.59 | g | 2.732E−02 | 0.9969 | 1.705E−02 |
| 1291.59 | 43.21 | g | 2.644E−02 | 0.9957 | 1.482E−02 |
| 1481.7 | 0.0603 | g | 2.555E−02 | 0.9938 | 2.297E−05 |
| | | | | $\Sigma =$ | 3.212E−02 |

The summation yields $3.212 \times 10^{-2}$ MeV cm$^2$ g$^{-1}$, and therefore the value of $\Gamma_\delta$ for $\delta = 5$ keV becomes 147.40 m$^2$ $\mu$Gy GBq$^{-1}$ h$^{-1}$.
In the case of $\delta = 20$ keV, where the first three x-ray lines have energies below the cut-off the summation yields $3.210 \times 10^{-2}$ MeV cm$^2$ g$^{-1}$ (note that the intensities of these x rays are rather low, and therefore their contributions are very small), and the value of $\Gamma_{\delta=20\, keV}$ is 147.32 m$^2$ $\mu$Gy GBq$^{-1}$ h$^{-1}$.
Compared with the value of 145.9 m$^2$ $\mu$Gy GBq$^{-1}$ h$^{-1}$ in Table 18.2 of the textbook, the difference is only 1% even if different disintegration scheme and $(\mu_{en}/\rho)_{air}$ data have been used.
The corresponding $\Gamma_X$ is simply obtained from Eq. (18.21) in the textbook, that is, for $\Gamma_{20}$,

$$\Gamma_X = \frac{\Gamma_{20}}{236.8} = 0.6221\, R\, m^2\, Ci^{-1}\, h^{-1}$$

**3** An old radiation therapy $^{60}$Co source was illegally abandoned in a scrapyard and then stolen and transported for 2 hours in a truck. The truck was subsequently abandoned in a field with the source inside. This had a label stating '7282 Ci on 1993-07-08' and it was estimated that the transport operation took place on 2013-11-20. Assuming that the source was situated at about 2 m from the driver and that the cabin wall thickness was ∼1 cm of iron, make an estimate of the dose received by the driver's body surface. *Answer: 2.27 Gy.*

**Solution:**

The solution involves a direct application of the air-kerma rate constant. As several values have been given in the text for $^{60}$Co, which vary with the energy cut-off ($\delta$), the radionuclide disintegration scheme, and mass energy-absorption coefficients, we start with an estimate of the air-kerma rate constant under a well-established set of data. We calculate $\Gamma$ for a cut-off $\delta$ of 5 keV, that is, $\Gamma_5$, using the $^{60}$Co disintegration scheme from the BIPM-LNHB (www.nucleide.org/DDEP_WG/DDEPdata.htm) and $\mu_{en}/\rho$ data from the Data Tables. Retrieving from the web site the text file 'Co-60 lara.txt', we extract the following information:

| Nuclide | Co-60 |
|---|---|
| Element | Cobalt |
| Z | 27 |
| Daughter(s) | $(\beta^-)$ Ni-60 |
| Half-life (a) | 5.2711 |
| Half-life (s) | 1.663E+08 |
| Decay constant (1/s) | 4.167E−09 |
| Mass activity (Bq/g) | 4.182E+13 |

and the disintegration scheme, from which we prepare the following table:

| Energy, $k_i$ (keV) | Intensity (%) 100 $n_i$ | Type | $[\mu_{en}/\rho(k_i)]_{air}$ (cm$^2$ g$^{-1}$) | $1 - \bar{g}_{air}$ | $k_i\, n_i\, \frac{[\mu_{en}(k_i)/\rho]_{air}}{1-\bar{g}_{air}}$ |
|---|---|---|---|---|---|
| 0.84 | 0.0002 | XL | 5.577E+03 | 1.0000 | 0.000E+00 |
| 7.46097 | 0.00334 | XKa2 | 1.140E+01 | 1.0000 | 2.841E−06 |
| 7.47824 | 0.0065 | XKa1 | 1.132E+01 | 1.0000 | 5.502E−06 |
| 8.2967 | 0.00136 | XKb1 | 8.230E+00 | 1.0000 | 9.287E−07 |
| 347.14 | 0.0075 | g | 2.916E−02 | 0.9993 | 7.598E−07 |
| 826.1 | 0.0076 | g | 2.869E−02 | 0.9980 | 1.805E−06 |
| 1173.228 | 99.85 | g | 2.698E−02 | 0.9966 | 3.172E−02 |
| 1332.492 | 99.9826 | g | 2.623E−02 | 0.9953 | 3.511E−02 |
| 2158.57 | 0.0012 | g | 2.285E−02 | 0.9835 | 6.019E−07 |
| 2505.692 | 0.000002 | g | 2.176E−02 | 0.9789 | 1.114E−09 |
| | | | | $\Sigma =$ | 6.68E−02 |
| | | | | $\Gamma = 4590 \times \Sigma =$ | 306.80 |

We therefore conclude that $\Gamma_5 = 306.80$ µGy m$^2$ GBq$^{-1}$ h$^{-1}$, which is quite close to the value given in Table 18.6 of the textbook.

We also make use of the disintegration scheme to estimate weighted mean values of the various photon coefficients that will be needed below. Recall that, as for the estimation of mean energies, averages can be obtained as fluence, energy-fluence, or air-kerma weighted mean values. With the exception of the $(\bar{\mu}_{en}/\rho)_{w,air}$ used in the last step, all the coefficients are needed for attenuation calculations in air and iron; thus they will be calculated as fluence (proportional to the number of disintegrations for each x ray or γ line) weighted averages, that is,

$$\bar{\mu} = \frac{\sum_i \mu_i \, n_i}{\sum_i n_i}$$

For $(\bar{\mu}_{en}/\rho)_{w,air}$, we should use an energy-fluence weighted calculation, to account for the energy transported by the beam (similar to what we do in the calculation of $\Gamma$), although in this particular case the fluence and energy-fluence weighted values coincide. The following table can thus be drawn up:

| $k_i$ (keV) | $(\mu/\rho)_{air}$ (cm$^2$ g$^{-1}$) | $1 - g_{air}$ | $(\mu_{en}/\rho)_{Fe}$ (cm$^2$ g$^{-1}$) | $(\mu/\rho)_{Fe}$ (cm$^2$ g$^{-1}$) | $(\mu_{en}/\rho)_{w}$ (cm$^2$ g$^{-1}$) | $(\mu_{en}/\rho)_{w,air}$ |
|---|---|---|---|---|---|---|
| 0.84 | 5.589E+03 | 1.0000 | 1.280E+04 | 1.285E+04 | 6.289E+03 | 1.1277 |
| 7.46097 | 1.193E+01 | 1.0000 | 2.686E+02 | 3.633E+02 | 1.200E+01 | 1.0523 |
| 7.47824 | 1.185E+01 | 1.0000 | 2.672E+02 | 3.611E+02 | 1.191E+01 | 1.0523 |
| 8.2967 | 8.697E+00 | 1.0000 | 2.115E+02 | 2.772E+02 | 8.632E+00 | 1.0488 |
| 347.14 | 1.010E−01 | 0.9993 | 3.157E−02 | 1.013E−01 | 3.240E−02 | 1.1111 |
| 826.1 | 6.963E−02 | 0.9980 | 2.696E−02 | 6.590E−02 | 3.189E−02 | 1.11115 |
| 1173.228 | 5.864E−02 | 0.9966 | 2.501E−02 | 5.521E−02 | 2.999E−02 | 1.1116 |
| 1332.492 | 5.496E−02 | 0.9953 | 2.426E−02 | 5.176E−02 | 2.914E−02 | 1.1107 |
| 2158.57 | 4.262E−02 | 0.9835 | 2.158E−02 | 4.122E−02 | 2.543E−02 | 1.1129 |
| 2505.692 | 3.935E−02 | 0.9789 | 2.095E−02 | 3.870E−02 | 2.424E−02 | 1.1142 |
| Weighted averages | 6.304E−02 | 0.9959 | 5.206E−02 | 8.604E−02 | | 1.1112 |

The second step is to estimate the source activity at the time of its transport. The number of days elapsed between 1993-07-08 and 2013-11-20 is estimated to be 7440, that is, 642 816 000 s. For an initial activity of 7282 Ci, this means that the activity on the transport date is estimated to be, see Eq. (18.6) in the textbook,

$$\mathcal{A} = \mathcal{A}_0 \, e^{-\lambda t} = 7282 \times e^{-1.663 \times 10^8 \times 642816000} \approx 500.0 \text{ Ci} = 1.850 \times 10^4 \text{ GBq}$$

Therefore the air-kerma rate in vacuum will be, see Eq. (18.22) in the textbook,

$$\dot{K}_{air}(0) = \Gamma_\delta \frac{A}{l^2} = 306.80 \times \frac{1.850 \times 10^4}{2^2}$$
$$= 1.419 \times 10^6 \ \mu Gy \ h^{-1} = 1.419 \ Gy \ h^{-1}$$

This figure will be attenuated by a distance of 199 cm of air, that is, 0.2397 g cm$^{-2}$ for $\rho_{air} = 1.2045 \times 10^{-3}$ g cm$^{-3}$, yielding

$$\dot{K}_{air}(t_{air}) = \dot{K}_{air}(0) \ e^{-\bar{\mu}_{air} \ t_{air}} = 1.419 \times e^{-0.06304 \times 0.2397} = 1.398 \ Gy \ h^{-1}$$

as well as by 1 cm of iron, that is, 7.8740 g cm$^{-2}$ for $\rho_{Fe} = 7.874$ g cm$^{-3}$, which gives

$$\dot{K}_{air}(t_{Fe}) = \dot{K}_{air}(t_{air}) \ e^{-\bar{\mu}_{Fe} \ t_{Fe}} = 1.398 \times e^{-0.08604 \times 7.8740} = 0.710 \ Gy \ h^{-1}$$

In addition, because the cabin wall is practically in contact with the driver's back, we need to account for the build-up it produces. We can make a rough estimate of it using Eq. (18.82) of the textbook, valid for low or medium atomic numbers ($Z_{Fe} = 26$), that is,

$$B(^{60}Co, t_{Fe}) \approx 1 + \frac{\bar{\mu}_{Fe} - \bar{\mu}_{en,Fe}}{\bar{\mu}_{en,Fe}} \bar{\mu}_{Fe} \ t_{Fe}$$
$$= 1 + \frac{8.604 - 5.206}{5.206} \times 8.604 \times 10^{-2} \times 7.8740 = 1.442$$

and therefore the air-kerma rate at the position of the driver's back becomes

$$\dot{K}_{air}(t_{Fe+buildup}) = \dot{K}_{air}(t_{Fe}) \ B(^{60}Co, t_{Fe}) = 0.710 \times 1.442 = 1.024 \ Gy \ h^{-1}$$

which is equivalent to a dose rate in air of

$$\dot{D}_{air} = \dot{K}_{air} \overline{(1 - g_{air})} = 1.024 \times 0.9959 = 1.020 \ Gy \ h^{-1}$$

The last step involves transferring $\dot{D}_{air}$ to $\dot{D}_{med}$. Assuming water to be approximately equivalent to tissue, we can write

$$\dot{D}_w = \dot{D}_{air} \ (\bar{\mu}_{en}/\rho)_{w,air} = 1.020 \times 1.111 = 1.133 \ Gy \ h^{-1}$$

which in 2 h corresponds to an absorbed dose of 2.27 Gy.
Note that all mass attenuation and energy-absorption coefficients are taken from the electronic Data Tables.

4  Determine the absorbed-dose rate in water at a distance of 2.5 mm from a 1.62 MBq point β source emitting 0.8 electrons per disintegration with an energy of 0.96 MeV.
   *Answer: 1 mGy s$^{-1}$.*

**Solution:**
The scaling parameter $X_{90}$ is approximated by (see text below Eq. (18.34) in the textbook)

$$X_{90} \approx 10^{-0.4846+1.2857 \ \log_{10}(E_0)-0.1924 \ [\log_{10}(E_0)]^2} = 0.31 \ cm$$

and the ratio $r/X_{90}$ becomes

$$r/X_{90} = \frac{0.25}{0.31} = 0.80$$

From Figure 18.9 in the textbook, we can estimate $F(E = 0.96 \text{ MeV}, r/X_{90} = 0.8) \approx 1.22$, and the specific absorbed-dose fraction results from Eq. (18.34)

$$AF_{\text{m}}(0.96, 0.8) = \frac{F(0.96, 0.8)}{4\pi r^2 \rho_{\text{water}} \ X_{90}} = \frac{1.22}{4\pi \times 0.25^2 \times 0.9982 \times 0.31} = 5.01 \text{ g}^{-1}$$

where we have used $\rho_{\text{water}} = 0.9982$ g cm$^{-3}$, that is, at room temperature. The absorbed-dose rate, from Eq. (18.35) in the textbook, therefore is

$$\dot{D} = A \, n \, E \, AF_{\text{m}}(E, r/X_{90})$$
$$= 1.62 \times 10^6 \text{ Bq} \times 0.8 \text{ Bq}^{-1}\text{s}^{-1} \times 0.96 \text{ MeV} \times 5.01 \text{ g}^{-1}$$
$$\times \frac{1.6022 \times 10^{-13} \text{ J MeV}^{-1}}{10^{-3} \text{ kg g}^{-1}}$$
$$\approx 10^{-3} \text{ Gy s}^{-1} = 1 \text{ mGy s}^{-1}$$

**5**  A water object contains a uniformly distributed γ-ray source. The volume of interest has a radius of 5 cm, and its center is inside the boundary of the object at an average distance $\bar{r} = 20$ cm from the boundary. (a) Estimate the absorbed dose at the object center for a total energy of $10^{-2}$ J converted in each kilogram of the object. (b) Compare the result for the 1 MeV γ rays with that obtained using Figure 18.7 in the textbook. (c) Use the straight-ahead approximation of Eq. (18.31) in the textbook to repeat estimate (a).
*Answer: (a) 4.81 × 10⁻³ Gy; (b) 2.72 × 10⁻³ Gy; (c) 4.51 × 10⁻³ Gy.*

**Solution:**
The linear attenuation and energy-absorption coefficients for the 1 MeV γ rays in water are $\mu = 0.0706$ cm$^{-1}$ and $\mu_{\text{en}} = 0.0310$ cm$^{-1}$, respectively. The mean free path therefore is MFP $= 1/\mu \approx 14$ cm.

(a) For a 5 cm radius water sphere (a mass of 0.524 kg) the total energy converted per kilogram is $5.236 \times 10^{-3}$ J kg$^{-1}$. Note that the 1 MeV photon MFP is smaller than the average distance of the center from the boundary of the object. Furthermore, the central point is far enough from the boundary so that the effect of backscattering is negligible, whether the object is surrounded by inert medium or by vacuum.

For 1 MeV γ rays and $\bar{r} = 20$ cm, from Table 18.3 in the textbook, the absorbed-dose fraction is $AF_{dv,V} = 0.518$. This is the approximate fraction of the γ-ray energy that contributes to the dose at the center, that is, the dose from these γ rays is $0.518 \times 5.236 \times 10^{-3}$ J kg$^{-1} = 2.713 \times 10^{-3}$ J kg$^{-1}$.

(b) For a very large object, where $AF_{dv,V} \approx 1$, the RE dose at the center would be $5.236 \times 10^{-3}$ J kg$^{-1}$.

If we had assumed the object to be a 20 cm radius water sphere, we could compare this result with Figure 18.7 of the textbook. It can be seen there that about 52% absorption of 1 MeV γ rays occurs for $\bar{r} = 20$ cm; therefore, the γ-ray dose at the center is this percentage of its RE value of

5.236 × 10$^{-3}$ Gy, that is, 2.723 × 10$^{-3}$ Gy, in good agreement (0.3%) with the value obtained in (a).

(c) Using the straight-ahead approximation of Eq. (18.31) in the textbook, the absorbed-dose fraction is $AF_{dv,V} \approx 1 - e^{-\mu_{en} \bar{r}} = 0.462$. It would yield a γ-ray dose of 2.420 × 10$^{-3}$ Gy instead of the 2.713 × 10$^{-3}$ Gy. This is approximately 12% difference, which reflects the level of accuracy usually provided by the straight-ahead approximation.

**6** What is the absorbed-dose rate (in Gy h$^{-1}$) at the center of a sphere of water 1 cm in radius, containing the β emitter $^{32}_{15}$P homogeneously distributed, with 6 × 10$^5$ disintegrations per second occurring per gram of water? The maximum energy of the $\beta^-$ rays from $^{32}_{15}$P is 1.711 MeV, and the mean energy of the beta spectrum is 0.696 MeV. (Assume constancy of disintegration rate and disregard electron capture transitions.)
*Answer: 0.24 Gy h$^{-1}$.*

**Solution:**
The range of electrons with the maximum energy of the $\beta^-$ rays from $^{32}_{15}$P is $R_{CSDA}(E_{\beta_{max}} = 1.711$ MeV$) \approx 8$ mm of water. Thus CPE will be established at the center of the 1 cm radius sphere.
The absorbed-dose rate there will be

$$\dot{D}_{\beta^-} = 3600\frac{s}{h} \times 6 \times 10^5 \frac{dis}{g\,s} \times 0.696\frac{MeV}{dis}$$

$$\times 1.6022 \times 10^{-10}\frac{Gy}{MeV/g} = 0.24 \text{ Gy h}^{-1}$$

**7** A sphere of water with a radius of 5 cm contains a uniform distribution of the $\beta^+$ emitter $^{15}_8$O (see disintegration scheme in Figure 18.13 of the textbook; $E_{\beta_{max}} = 1.735$ MeV, $\bar{E}_\beta = 0.737$ MeV), with a level of activity of 10$^6$ dis g$^{-1}$ s$^{-1}$. (a) What is the approximate absorbed-dose rate (in Gy h$^{-1}$) at the center of the sphere, estimated by the mean radius ($\bar{r}$) straight-ahead approximation, see Eq. (18.31). (b) What would the answer to (a) become if the radioactive water sphere were increased to a radius of 150 cm?
*Answer: (a) 0.515 Gy h$^{-1}$; (b) 1.010 Gy h$^{-1}$.*

**Solution:**
(a) CPE exists at the center, as $R_{CSDA}(E_{\beta_{max}} = 1.735$ MeV$) \approx 8$ mm water. The $\beta^+$ absorbed-dose rate will thus be

$$\dot{D}_{\beta^+} = 3600\frac{s}{h} \times 10^6 \frac{dis}{g\,s} \times 0.737\frac{MeV}{dis} \times 1.6022$$

$$\times 10^{-10}\frac{Gy}{MeV/g} = 0.425 \text{ Gy h}^{-1}$$

For each disintegration producing one $\beta^+$ particle, two 0.511 MeV γ rays are produced when the particle stops. For these γ rays, $\mu_{en} = 0.033$ cm$^{-1}$. Using the straight-ahead approximation, the absorbed-dose fraction of the energy per unit mass emitted as $\gamma$ rays that contributes to the dose is AF $\approx 1 - e^{-\mu_{en}\bar{r}} = 0.152$. Thus the dose rate contributed by $\gamma$ rays is

$$\dot{D}_\gamma = 3600 \, \frac{s}{h} \times 10^6 \, \frac{dis}{g\,s} \times 1.022 \, \frac{\gamma\text{-ray MeV produced}}{dis}$$

$$\times 0.152 \, \frac{\gamma\text{-energy absorbed}}{\gamma\text{-energy produced}} \times 1.6022$$

$$\times 10^{-10} \, \frac{Gy}{MeV/g} = 0.090 \, Gy\,h^{-1}$$

Therefore, the total absorbed-dose rate at the sphere center, including the $\beta^+$ contribution, is

$$\dot{D}_{tot} = \dot{D}_{\beta^+} + \dot{D}_\gamma = 0.425 + 0.090 = 0.515 \, Gy\,h^{-1}$$

(b) Increasing $\bar{r}$ to 150 cm and using the straight-ahead approximation increases the absorbed-dose fraction to 0.993. The $\gamma$-ray dose rate contribution at the center becomes 0.585 $Gy\,h^{-1}$, and the total dose rate is $\dot{D}_{tot} = 0.425 + 0.585 = 1.010 \, Gy\,h^{-1}$.

Note that since the sphere radius in (b) is so large ($\approx 14$ MFP), RE should be closely approximated at the center. This means that the atomic rest mass decrease should nearly all appear as dose except for what escapes with neutrinos. From the disintegration scheme in Figure 18.13 of the textbook, it can be seen that the total energy derived from rest mass per disintegration of $^{15}_{8}O \rightarrow ^{15}_{7}N$ is 2.757 MeV, while the amount of energy carried away by neutrinos can be approximated by $1.735 - 0.737 = 0.998$ MeV. Thus the energy going into dose under RE conditions is simply 2.757 MeV $- 0.998$ MeV $= 1.759$ MeV per disintegration. Therefore, the dose rate assuming RE is

$$\dot{D}_{tot} = 3600\frac{s}{h} \times 10^6\frac{dis}{g\,s} \times 1.759\frac{MeV}{dis} \times 1.6022 \times 10^{-10}\frac{Gy}{MeV/g}$$

$$= 1.015 \, Gy\,h^{-1}$$

which is practically the same as that obtained in (b), as it should be.

**8** A water sphere of 2 cm radius contains a uniformly distributed source of $^{22}_{11}Na$ (see the $\beta^+$ and electron capture disintegration scheme in Figure 18.14 of the textbook; $E_{\beta_{max}} = 0.546$ MeV, $\bar{E}_\beta = 0.216$ MeV), which undergoes $10^5$ dis $g^{-1}\,s^{-1}$. Estimate the absorbed dose deposited at its center in one week, using the mean-radius, straight-ahead approximation. Compare the result with that given in Table 18.4 of the textbook.
*Answer: 3.167 Gy. The difference of about 0.6% from the result in Table 18.4 is due to the use of the straight-ahead approximation for the absorbed-dose fraction of the two types of $\gamma$ rays involved.*

**Solution:**
From the disintegration scheme in Figure 18.14, it can be seen that there are 90.3% $\beta^+$ events per disintegration and 9.64% EC events per disintegration.

For the $\beta^+$ particles, CPE exists because of the small $R_{CSDA}(E_{\beta_{max}} = 0.546 \text{ MeV})$ in water, which is approximately 2 mm. Thus the dose they deposit, for $0.903 \ \beta^+/\text{dis}$ and $\bar{E}_{\beta} = 0.216 \text{ MeV}$, is

$$D_{\beta^+} = 10^5 \frac{\text{dis}}{\text{g s}} \times 0.903 \frac{\beta^+}{\text{dis}} \times 0.216 \frac{\text{MeV}}{\beta^+} \times 1.6022 \times 10^{-10} \frac{\text{Gy}}{\text{MeV/g}}$$
$$\times 6.048 \times 10^5 \text{ s} = 1.887 \text{ Gy}$$

The very small dose contribution due to the $U_B$-term for the EC events may be calculated as follows, although here, as in most cases, it will be found to be negligible unless the competing $\beta^+$ transitions are forbidden by energy considerations. All of the binding energy of the electrons captured in the parent nuclei of $^{22}_{11}\text{Na}$ is given either to Auger electrons or to fluorescence photons. As shown in the Data Tables, K-fluorescence photons from the daughter atom Ne have a very low energy (actually 0.87 keV); moreover, the fluorescence yield for Ne, $\omega_K = \omega_L \approx 0$, which means that the $U_B$ of the parent-atom electrons is entirely absorbed in the immediate vicinity of the atom in question. For the K-shell of sodium $(U_B)_K = 1.075 \text{ keV}$, while $(U_B)_{L1} = 0.066 \text{ keV}$. The contribution from the $U_B$-term for EC events can then be approximated by

$$f_{EC} \approx 0.9(U_B)_K + 0.1(U_B)_L = 9.741 \times 10^{-4} \frac{\text{MeV}}{\text{EC event}}$$

where 90% of the EC events involve the K-shell and 10% can be assumed to be with the L-shell (see photo-fraction in Data Tables). Considering that there are 0.964 EC events per disintegration, the dose contributed by EC events is

$$D_{EC} = 10^5 \frac{\text{dis}}{\text{g s}} \times 0.0964 \frac{\text{EC events}}{\text{dis}} \times f_{EC} \frac{\text{MeV}}{\text{EC event}} \times 1.6022$$
$$\times 10^{-10} \frac{\text{Gy}}{\text{MeV/g}} \times 6.048 \times 10^5 \text{ s} = 9.10 \times 10^{-4} \text{ Gy}$$

which compared with $D_{\beta^+}$ is, of course, negligible.

Next we must calculate the $\gamma$-ray contribution to the dose at the center of the sphere, for which $\bar{r} = 2 \text{ cm}$. For each beta-plus disintegration, 0.903 $\beta^+$ particles are emitted on the average; consequently $2 \times 0.903 = 1.806$ photons of 0.511 MeV each are emitted, for which $\mu_{en} = 0.033 \text{ cm}^{-1}$ and thus the absorbed-dose fraction estimated by the straight-ahead approximation (Eq. 18.31) is AF $\approx 1 - e^{-\mu_{en} \bar{r}} = 0.064$. Therefore, the resulting dose is

$$D_{\gamma 0.511} = 10^5 \frac{\text{dis}}{\text{g s}} \times 1.806 \frac{\text{photons}}{\text{dis}} \times 0.511 \frac{\text{MeV}}{\text{photon}} \times 1.6022$$
$$\times 10^{-10} \frac{\text{Gy}}{\text{MeV/g}} \times 6.048 \times 10^5 \text{ s} \times 0.064 = 0.570 \text{ Gy}$$

For the final relaxation $\gamma$ ray in the disintegration scheme from $^{22}_{10}\text{Ne}^* \rightarrow ^{22}_{10}\text{Ne}$ of approximately 1.275 MeV, which occurs for every disintegration,

$\mu_{en} = 0.0295 \text{ cm}^{-1}$ and therefore AF $= 0.057$; thus the dose is

$$D_{\gamma 1.275} = 10^5 \frac{\text{dis}}{\text{g s}} \times 1 \frac{\text{photon}}{\text{dis}} \times 1.275 \frac{\text{MeV}}{\text{photon}} \times 1.6022$$
$$\times 10^{-10} \frac{\text{Gy}}{\text{MeV/g}} \times 6.048 \times 10^5 \text{ s} \times 0.057 = 0.709 \text{ Gy}$$

Hence the total dose is

$$D_{tot} = D_{\beta^+} + D_{EC} + D_{\gamma 0.511} + D_{\gamma 1.275}$$
$$= 1.887 + 9.10 \times 10^{-4} + 0.570 + 0.709 = 3.167 \text{ Gy}$$

of which about 60% is from the $\beta^+$ component.

Comparing this result with that given in Table 18.4 of the textbook, 3.147 Gy, the difference is about 0.6%. This is due to the use, in the present exercise, of the straight-ahead approximation for the absorbed-dose fraction of the two types of $\gamma$ rays involved, whereas in Table 18.4 the values were taken from the Monte Carlo data in Table 18.3 of the textbook.

**9** A sphere of water 10 cm in diameter contains a uniform source of $^{137}$Cs (see the disintegration scheme in Figure 18.15 of the textbook) undergoing $10^3$ dis g$^{-1}$ s$^{-1}$. What is the absorbed dose at the center for a 10-day period, due only to the decay of $^{137m}_{56}$Ba? For the absorbed-dose fraction of $\gamma$ rays, compare the straight-ahead approximation with the value from Table 18.3 in the textbook and, based on the difference obtained, make a choice for the AF value to be used.

*Answer: 2.081 ×10$^{-2}$ Gy, using AF data from Table 18.3, which differs by ∼ 5% from the straight-ahead approximation.*

**Solution:**

For the $\gamma$ rays of 0.662 MeV (with an emission of ∼ 85%), Table 18.3 yields (interpolating)

$$\text{AF}(0.662 \text{ MeV}, 5 \text{ cm}) = 0.1582$$

whereas the straight-ahead approximation, with $\mu_{en} \approx 0.03254 \text{ cm}^{-1}$, gives

$$\text{AF}(0.662 \text{ MeV}, 5 \text{ cm}) \approx 1 - e^{-0.0326 \times 5} = 0.1502$$

that is, a difference of ∼ 5.4%; we will therefore use the more accurate data from Table 18.3.

The $\gamma$-rays dose in 10 days then is

$$D_\gamma = 10^3 \frac{\text{dis}}{\text{g s}} \times 0.85 \frac{\gamma \text{ rays}}{\text{dis}} \times 0.662 \frac{\text{MeV}}{\gamma \text{ ray}} \times 1.6022$$
$$\times 10^{-10} \frac{\text{Gy}}{\text{MeV/g}} \times 8.64 \times 10^5 \text{ s} \times 0.1582 = 1.235 \times 10^{-2} \text{ Gy}$$

For the K-shell conversion process (7.62%, see text related to Figure 18.15), making use of Eq. (18.55), the dose contribution will be

$$D_{IC}^{K} = 10^3 \frac{dis}{g\,s} \times 0.076 \frac{IC(K)\,\text{events}}{dis}$$

$$\times (k - p_K \omega_K \bar{k}_K) \frac{MeV}{IC(K)} \times 1.6022 \times 10^{-10} \frac{Gy}{MeV/g}$$

$$\times 8.64 \times 10^5 \text{ s} = 1.055 \times (0.662 - p_K\,\omega_K\,\bar{k}_K) \times 10^{-2} \text{ Gy}$$

From the Data Tables, the mean K-shell fluorescent energy is $\bar{k}_K = 32.9$ keV and the fluorescent yield $\omega_K \approx 0.90$, with $\mu_{en} = 0.1166$ cm$^{-1}$ (for 0.0329 MeV). Hence Eq. (18.56) gives $p_K = e^{-0.1166 \times 5} = 0.559$ and therefore

$$D_{IC}^{K} = 6.811 \times 10^{-3} \text{ Gy}$$

Likewise, for the IC process in the $L + M + \cdots$ shells (1.75%, see text related to Fig 18.15), which we may assume to be all $L_1$-shell, $\bar{k}_L \approx 5$ keV, for which $\mu_{en} = 23.44$ cm$^{-1}$; thus $p_L \approx 0$, and the contribution is

$$D_{IC}^{L} = 10^3 \frac{dis}{g\,s} \times 0.0175 \frac{IC(L)\text{events}}{dis} \times (0.662 - 0) \frac{MeV}{IC(L)}$$

$$\times 1.6022 \times 10^{-10} \frac{Gy}{MeV/g} \times 8.64 \times 10^5 \text{ s} = 1.656 \times 10^{-3} \text{ Gy}$$

Hence the total absorbed dose in 10 days due to the disintegration of $^{137m}_{56}$Ba atoms is

$$D_{tot} = D_\gamma + D_{IC}^{K} + D_{IC}^{L} = 1.235 \times 10^{-2} + 6.811 \times 10^{-3} + 1.656 \times 10^{-3}$$

$$= 2.081 \times 10^{-2} \text{ Gy}$$

**10**  20 MBq of $^{90}$Y are administered to a tumor (assumed to be a water-equivalent sphere) with a volume of 20 ml. Assuming that the radionuclide is homogeneously distributed within the tumor and has an infinite biological half-life, calculate the absorbed dose to the tumor 24 h after the administration of the radionuclide. The absorbed-dose fraction evaluated at the mean beta energy is 0.9093 (Stabin and Konijnenberg, 2000). If possible, use electron emissions from the disintegration scheme in www.nucleide.org/DDEP_WG/DDEPdata.htm.
*Answer: 10.27 Gy.*

**Solution:**
From the Internet site, we get $T_{1/2} = 2.6684$ days, and

| n (Electrons) per 100 disintegration | Average energy (keV) |
| --- | --- |
| $1.4 \times 10^{-6}$ | 24.5 |
| 0.017 | 163.7 |
| 99.983 | 926.7 |

We use $T_{1/2} = 64.04$ h, yielding $\lambda = \ln 2/T_{1/2} = 0.0108$ h$^{-1}$, which is taken together with the rest of the input data, that is, $A_0 = 20$ MBq; $V = 20$ cm$^3$; $T_D = 24$ h; AF (absorbed-dose fraction, $\varphi$) $= 0.9093$ and, as the biological half-life is infinite, $\lambda_b = \ln 2/T_b = \ln 2/\infty = 0$.

The mean absorbed dose over a time period $T_D$ can be approximated by (see Eq. (18.63))

$$\bar{D} = \frac{\tilde{\mathcal{A}}}{m} \sum_i n_i E_i \varphi_i \approx \frac{\tilde{\mathcal{A}}}{m} \varphi(\bar{E}) \sum_i n_i E_i$$

where $\tilde{\mathcal{A}}$ is the time-integrated activity during the time period $T_D$ considered, $m$ is the mass of the target region, $\varphi(\bar{E})$ is the absorbed-dose fraction (dimensionless) that, for simplicity, has been approximated to the value at the overall mean energy of the set of beta-energies (average)and given as input data, and $n_i$ and $E_i$ are the emission data given in the table above.

The time-integrated activity is calculated as (see Eq. (18.60))

$$\tilde{\mathcal{A}}(T_D) = \int_0^{T_D} \tilde{A}(t) \, e^{-(\lambda + \lambda_b) t} \, dt$$

$$= \frac{A_0}{\lambda} [e^{-\lambda t}]_0^{24\text{h}} = \frac{20 \text{ MBq}}{0.0108 \text{ h}^{-1}} (1 - 0.772) = 422.72 \text{ MBq h}$$

which considering that 1 MBq $= 10^6$ Bq and h $= 3600$ s, yields

$$\tilde{\mathcal{A}} = 1.5218 \times 10^{12} \text{ Bq s} = 1.5218 \times 10^{12}$$

and as Bq $= $ s$^{-1}$, $\tilde{\mathcal{A}}$ is dimensionless.

The mass of the water-equivalent 20 cm$^3$ sphere is $m = \rho V = 1.0 \text{ g cm}^{-3} \times 20 \text{ cm}^3 = 20$ g.

The summation in $\bar{D}$ is calculated as

$$\sum_i n_i E_i = 24.5 \text{ keV} \times \frac{1.4 \times 10^{-6}}{100} + 163.7 \text{ keV}$$

$$\times \frac{0.017}{100} + 926.7 \text{ keV} \times \frac{99.983}{100} = 926.57 \text{ keV}$$

(Note that this is the mean energy, used to calculate the absorbed-dose fraction, given as input data.)

We therefore calculate the mean absorbed dose from

$$\bar{D} = \frac{1.5218 \times 10^{12}}{20 \text{ g}} \times 0.9093 \times 926.57 \text{ keV} = 6.4108 \times 10^{13} \text{ keV g}^{-1}$$

which, using the conversions 1 keV $= 1.6022 \times 10^{-16}$ J and 1 g $= 10^{-3}$ kg, becomes

$$\bar{D} = 10.27 \text{ Gy}$$

For comparison, for a 20 g sphere and 422.7 MBq h, the OLINDA software gives 10.3 Gy.

**11** Determine the apparent activity of a $^{125}$I brachytherapy seed whose air-kerma rate at 1 m is 0.526 µGy h$^{-1}$.
*Answer: 15 MBq.*

**Solution:**
The expression for the apparent activity is, see Eq. (18.84),

$$\mathcal{A}_{\text{app}} = \frac{\dot{K}_R \, r^2}{\Gamma_\delta} = \frac{S_K}{\Gamma_\delta}$$

where the air-kerma strength is

$$S_K = \dot{K}_R \, r^2 = 0.51 \ \text{µGy h}^{-1} \text{m}^2$$

From Table 18.6 in the textbook, the air-kerma rate constant is $\Gamma_5 = \Gamma_{10} = 34.03$ µGy m$^2$ GBq$^{-1}$ h$^{-1}$. Note that Table 18.2 gives $\Gamma_{20} = 37.73$, but this value is for a 'naked' source; in the present case, the value for a brachytherapy source should be used.
Therefore, the apparent activity is

$$\mathcal{A}_{\text{app}} = \frac{0.51 \ \text{µGy h}^{-1}\text{m}^2}{34.03 \ \text{µGy m}^2 \ \text{GBq}^{-1} \ \text{h}^{-1}} = 1.499 \ \text{GBq} = 15 \ \text{MBq}$$

**12** The calibration certificate of a $^{192}$Ir brachytherapy source *mHDR-v2 (Nucletron)* states a reference air-kerma rate of 46.32 mGy h$^{-1}$ at 1 m. The source has a length of 3.60 mm and a diameter of 0.65 mm. Calculate the absorbed-dose rate in water (in Gy min$^{-1}$) at a point at $r = 2$ cm and $\theta = 30°$ using the AAPM TG-43 formalism. Whenever possible, use data from AAPM-ESTRO Report 229, Perez-Calatayud *et al.* (2012), or from http://www.uv.es/braphyqs.
*Answer: 2 Gy min$^{-1}$.*

**Solution:**
The TG-43 expression for the absorbed-dose rate in water, Eq. (18.95), is

$$\dot{D}(r,\theta) = S_K \Lambda \, \frac{G_L(r,\theta)}{G_L(r_0,\theta_0)} \, g_L(r) \, F(r,\theta)$$

where
− $S_K$ is the air-kerma strength (mGy h$^{-1}$ cm$^2$),
− $\Lambda$ is the dose-rate constant[1] [mGy h$^{-1}$ $U^{-1}$], with $U$ in (mGy h$^{-1}$ cm$^2$),
− $G_L(r,\theta)$ is the geometry function (cm$^{-2}$),
− $g_L(r)$ is the radial-dose function (adimensional),
− $F(r,\theta)$ is the anisotropy function (adimensional), and
− $(r_0,\theta_0)$ is the reference point, at (1 cm, $\pi/2$).

---

1 Note that it is common to tabulate $\Lambda$ in (cGy h$^{-1}$ $U^{-1}$), with $U$ (cGy h$^{-1}$ cm$^2$), but the cancellation of units yields $\Lambda$ in (cm$^{-2}$) irrespective of the units used for absorbed dose and time (provided they are the same in $\Lambda$ and $U$). Note also that Table 18.6 provides a calculated value of $\Lambda = 1.035$, which is only valid for 'naked' sources; the value in AAPM Report 229 used here is for a specific brachytherapy source, taking filtration into account.

The numerical values for these quantities are

- $S_K = \dot{K}_R \, d_{\mathrm{ref}}^2 = 46.32 \text{ mGy h}^{-1} \times 100^2 \text{ cm}^2 = 463 \, 200 \text{ mGy h}^{-1}\text{cm}^2$
- $\Lambda$, $g_L(r)$, and $F(r, \theta)$ for this particular source are obtained from http://www.uv.es/braphyqs:
  $\Lambda = 1.109 \text{ mGy h}^{-1} \, U^{-1} = 1.109 \text{ mGy h}^{-1} \, (\text{mGy h}^{-1} \text{cm}^2)^{-1} = 1.109 \text{ cm}^{-2}$
  $g_L(r) = 1.005$
  $F(r, \theta) = 0.915$
- The geometry function is defined as (see Eq. (18.96))

$$G_L(r, \theta) = \frac{\varphi}{L \, r \, \sin \theta}$$

where the meaning of the different terms is illustrated in Figure 18.1 below. From the figure, using trigonometric relations, we get

$$\varphi = \theta_2 - \theta_1$$

$$\theta_1 = \mathrm{acos}\left( \frac{2r \cos \theta + L}{\sqrt{L^2 + 4r^2 + 4Lr \cos \theta}} \right)$$

$$\theta_2 = \mathrm{acos}\left( \frac{2r \cos \theta - L}{\sqrt{L^2 + 4r^2 - 4Lr \cos \theta}} \right)$$

that is, using $r = 2$ cm, $\theta = 30°$ and $L = 0.36$ cm, we obtain $\theta_1 = 0.4819$ rad and $\theta_2 = 0.5724$ rad, from where $\varphi = 0.0905$ rad. The geometry function value is therefore

$$G_L(2 \text{ cm}, 30°) = \frac{0.0905}{0.36 \text{ cm} \times 2 \text{ cm} \times \sin 30°} = 0.2514 \text{ cm}^{-2}$$

- Proceeding in a similar manner, for $G_L(r_0, \theta_0)$ we obtain $G_L(1 \text{ cm}, 90°) = 0.9894 \text{ cm}^{-2}$, with $\theta_1 = 1.3927$ rad and $\theta_2 = 1.7489$ rad yielding $\varphi = 0.3562$ rad.

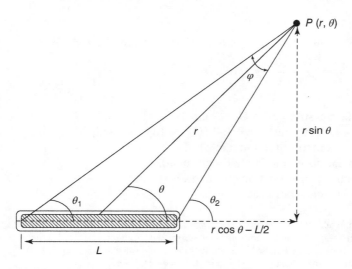

**Figure 18.1** Geometry used in the AAPM TG-43 formalism.

The absorbed-dose rate to water therefore is

$$\dot{D}(r,\theta) = 463\ 200\ \frac{\text{mGy cm}^2}{\text{h}} \times 1.109\ \frac{1}{\text{cm}^2} \times \frac{0.2514\ \text{cm}^2}{0.9894\ \text{cm}^2} \times 1.005 \times 0.915$$

$$= 120\ 005\ \text{mGy/h} = 2.000\ \text{Gy/min}$$

Note that, alternatively, the angles $\theta_1$ and $\theta_2$ could have been expressed as

$$\theta_1 = \text{atan}\left(\frac{r\sin\theta}{r\cos\theta + \frac{L}{2}}\right)$$

$$\theta_2 = \text{atan}\left(\frac{r\sin\theta}{r\cos\theta - \frac{L}{2}}\right)$$

but, as pointed out in AAPM Report 229, the use of the trigonometric function *atan* yields negative angles (instead of between 90° and 180°) for $\theta_2(90°) = -1.3927$, and therefore for $\varphi(90°) = -2.7854$, producing a negative value of $G_L(r_0, \theta_0) = G_L(1\ \text{cm}, 90°) = -7.7372\ \text{cm}^{-2}$.

# 19

## Neutron Dosimetry

1   Assume that the average neutron energy from a p(62) + Be(36) generator is
20 MeV and the neutrons interact with adipose tissue. Estimate the average
distance over which the C, N, and O recoil nuclei will deposit energy after
the neutrons have undergone elastic scattering.
*Answer: C: 41.8 µm; N: 31.4 µm; O: 26.3 µm.*

**Solution:**
The average distance can be approximated by the $R_{CSDA}$ corresponding to the
mean energies of the recoil nuclei. Hence, using Eq. (19.2) in the textbook for
the average energy transferred to a nucleus,

$$\bar{Q} = E_n \frac{2\,m_n\,M}{(m_n + M)^2}$$

for $E_n = 20$ MeV, we obtain

$$\bar{Q}(C\text{-}12) = 2.84 \text{ MeV}$$
$$\bar{Q}(N\text{-}14) = 2.49 \text{ MeV}$$
$$\bar{Q}(O\text{-}16) = 2.21 \text{ MeV}$$

(note that $M = A$ and $m_n = 1$ have been used, as in the neutron exercises
of Chapter 3). Using the Data Tables for the $R_{CSDA}$ of ions in water, noting
that the specific energies $E/A$ must be used ($\bar{Q}/A$ in this case), and scaling
$(R_{CSDA})_w$ by the adipose tissue density, $\rho_{adip} = 0.920$ g cm$^{-3}$, we obtain

| Nucleus | $\bar{Q}$ (MeV) | A | $\bar{Q}/A$ (MeV) | $(R_{CSDA})_w$ (cm) | $(R_{CSDA})_{adip}$ (µm) |
|---------|------|----|------|----------|----------|
| C | 2.84 | 12 | 0.237 | 3.84E−04 | 4.18 |
| N | 2.49 | 14 | 0.178 | 2.89E−04 | 3.14 |
| O | 2.21 | 16 | 0.138 | 2.42E−04 | 2.63 |

2   The fluence of thermal neutrons is $6.5 \times 10^{12}$ n cm$^{-2}$ in a layer of ICRU striated muscle 0.1 g cm$^{-2}$ thick. What is the absorbed dose at the center of the
layer?
*Answer: 1.81 Gy.*

*Fundamentals of Ionizing Radiation Dosimetry: Solutions to Exercises,* First Edition.
Pedro Andreo, David T. Burns, Alan E. Nahum, and Jan Seuntjens.
© 2017 Wiley-VCH Verlag GmbH & Co. KGaA. Published 2017 by Wiley-VCH Verlag GmbH & Co. KGaA.

**Solution:**
Combining Eq. (4.68) in the textbook for the neutron kerma, $K = \Phi \, k_\Phi$, with Eq. (19.1), the neutron kerma can be written as

$$K = C \, \Phi \, N \, \bar{Q} \, \sigma_{\text{th}}$$

where
- $C = 1.6022 \times 10^{-13}$ J MeV$^{-1}$
- $\Phi = 6.5 \times 10^{12}$ n cm$^{-2}$ = $6.5 \times 10^{16}$ n m$^{-2}$
- For $^{14}$N(n, p)$^{14}$C interactions (see text following Eq. (19.1)):
  $N = 1.504 \times 10^{24}$ atom/kg
  $\bar{Q} = 0.626$ MeV n$^{-1}$
  $\sigma = 1.827 \times 10^{-28}$ m$^2$/atom

The neutron dose then is

$$D_n \stackrel{\text{CPE}}{=} K_n = 1.6022 \times 10^{-13} \times 6.5 \times 10^{16} \times 1.504 \times 10^{24} \times 0.626$$
$$\times 1.827 \times 10^{-28} = 1.7914 \text{ Gy}$$

This is certainly the dominant dose contribution, but what about the $^1$H$(n, \gamma)^2$H interactions? The 2.225 MeV $\gamma$ rays mostly escape, and those electrons that they generate also mostly escape. This contribution can be estimated as follows:
From the text, $k/\Phi = 7.212 \times 10^{-16}$ Gy m$^2$. Hence, the energy converted to 2.225 MeV $\gamma$ rays per unit mass is

$$\frac{R_\gamma}{m} = 7.1212 \times 10^{-16} \frac{\text{J m}^2}{\text{kg n}} \times 6.5 \times 10^{16} \frac{\text{n}}{\text{m}^2} = 46.878 \frac{\text{J}}{\text{kg}}$$

Assuming the muscle layer to be infinite in lateral extent, the mean escape path length from the midplane is 1 mm, as the mean chord length through the 1 mm thick slab is $\bar{\ell} = 4 \, v/a = 2$ mm. Thus, the absorbed dose fraction (AF) of the 2.225 MeV $\gamma$ rays, for ICRU striated muscle with $\rho = 1.04$ g cm$^{-3}$ and $\mu_{\text{en}}/\rho = 0.0249$ cm$^2$ g$^{-1}$, is

$$\text{AF} = 1 - e^{-(\mu_{\text{en}}/\rho) \times 0.1 \text{ cm} \times 1.04 \text{ g cm}^{-3}}$$
$$= 1 - e^{-0.0249 \text{ cm}^2 \text{ g}^{-1} \times 0.104 \text{ g cm}^{-2}} = 2.586 \times 10^{-3}$$

Hence, the energy transferred to electrons by the $\gamma$ rays as they escape from the slab is

$$K_{\text{el}} = 46.878 \text{ J kg}^{-1} \times 2.586 \times 10^{-3} = 0.121 \text{ Gy}$$

However, CPE is not satisfied for these electrons for this thin layer, hence $D < K_{\text{el}}$. The average starting energy of the electrons is (assuming Compton interactions)

$$\bar{E}_0 = k \frac{\sigma_{\text{tr}}}{\sigma} = 2.225 \text{ MeV} \frac{7.456 \times 10^{-26}}{1.374 \times 10^{-25}} = 1.207 \text{ MeV}$$

for which, from the electronic Data Tables, $(S_{\text{el}}/\rho)_{\text{muscle}} = 1.8123$ MeV cm$^2$ g$^{-1}$. Therefore, the energy spent by electrons in escaping is approximately

(assuming constant stopping power)

$$1.8123 \text{ MeV} \frac{\text{cm}^2}{\text{g}} \times 0.104 \frac{\text{g}}{\text{cm}^2} = 0.1885 \text{ MeV}$$

or 0.156 of their starting energy. Hence

$$D_{\text{elec}} = 0.156 \, K_{\text{el}} = 0.156 \times 0.121 \text{ Gy} = 0.019 \text{ Gy}$$

and the total dose becomes

$$D_{\text{tot}} = D_{\text{n}} + D_{\text{elec}} = 1.7914 + 0.019 = 1.810 \text{ Gy}$$

**3**  Assume that the neutron fluence in the previous exercise is approximately uniform throughout a tissue (ICRU striated muscle) sphere 5 cm in radius, due to thermalization of a field of intermediate-energy neutrons penetrating the sphere. What is the absorbed dose at the center resulting from thermal-neutron-capture $\gamma$ rays? How does it compare with the total absorbed dose that results (directly or indirectly) from the thermal neutrons? *Answer: 5.69 Gy, 76% of total.*

**Solution:**
In this case there is CPE, therefore we can write

$$D_{\text{n}} \stackrel{\text{CPE}}{=} K_{\text{n}} = 1.791 \text{ Gy}$$

$$\frac{R_{\gamma}}{m} = 46.878 \frac{\text{J}}{\text{kg}}$$

Now the $\gamma$-absorbed dose fraction is

$$\text{AF} = 1 - e^{-(\mu_{\text{en}}/\rho) \times 5 \text{ cm} \times 1.04 \text{ g cm}^{-3}}$$
$$= 1 - e^{-0.0249 \text{ cm}^2 \text{ g}^{-1} \times 5.2 \text{ g cm}^{-2}} = 0.1214$$

Thus,

$$D_{\gamma} \stackrel{\text{CPE}}{=} (K_{\text{el}})_{\gamma} = 0.1214 \times 46.878 = 5.693 \text{ Gy}$$

and the total dose becomes

$$D_{\text{tot}} = D_{\text{n}} + D_{\gamma} = 1.791 + 5.693 = 7.485 \text{ Gy}$$

The gamma dose fraction is $5.693/7.485 = 0.76$, that is, 76% of the total dose is due to thermal-neutron-capture $\gamma$ rays.

**4**  Measurements in water with two detectors (an A-150-walled ionization chamber filled with TE-gas and a GM counter) in a p(62) + Be(36) neutron beam and a $^{60}$Co $\gamma$-ray beam yield a ratio $k_{\text{GM}}/k_{\text{A-150}} = 0.061$. The response-normalized readings for the two detectors are 12.52 Gy (chamber) and 2.82 Gy (GM). Calculate the total n + $\gamma$ absorbed dose assuming $h = 1$ for both detectors. What percentage of the total dose corresponds to $D_{\gamma}$? *Answer: 12.76 Gy, 17% of the total dose.*

**Solution:**
Using Eq. (19.5) in the textbook, assuming $h_{n+\gamma} = h_{ni} = 1$, and noting that $k_{n+\gamma} = k_{A\text{-}150}$ and $k_{ni} = k_{GM}$ the neutron and $\gamma$ rays absorbed doses are given by

$$D_n = \frac{M^R_{A\text{-}150} - M^R_{GM}}{k_{A\text{-}150} - k_{GM}}$$

$$D_\gamma = \frac{k_{A\text{-}150}\, M^R_{GM} - k_{GM}\, M^R_{A\text{-}150}}{k_{A\text{-}150} - k_{GM}}$$

where $M^R_{A\text{-}150} = 12.52$ Gy, $M^R_{GM} = 2.82$ Gy.
The value of $k_{A\text{-}150}$ can be determined from Eq. (19.12) in the textbook using the data in Table 19.3, i.e.,

$$k_{A\text{-}150} = \frac{(W_{TE\text{-gas}})_{60\,Co}}{(W_{TE\text{-gas}})_{n+\gamma}} \frac{(s_{A\text{-}150,TE\text{-gas}})_{60\,Co}}{(s_{A\text{-}150,TE\text{-gas}})_{n+\gamma}} \frac{(K_{w,\,A\text{-}150})_{60\,Co}}{(K_{w,\,A\text{-}150})_{n+\gamma}}$$

$$= \frac{29.3\ \text{eV}}{31.1\ \text{eV}} \times \frac{0.991}{1.0} \times \frac{1.012}{0.967} = 0.977$$

where $(K_{w,\,A\text{-}150})_{60\,Co}$ has been replaced by $[(\bar{\mu}_{en}/\rho)_{w,\,A\text{-}150}]_{60\,Co}$. Using $k_{GM}/k_{A\text{-}150} = 0.061$ results in

$$k_{GM} = 0.061 \times 0.977 = 0.060$$

The numerical values can then be inserted into the absorbed dose expressions above to yield

$$D_n = \frac{12.52 - 2.82}{0.977 - 0.060} = 10.57\ \text{Gy}$$

$$D_\gamma = \frac{0.977 \times 2.82 - 0.060 \times 12.52}{0.977 - 0.060} = 2.19\ \text{Gy}$$

Hence

$$D_{tot} = D_n + D_\gamma = 10.57 + 2.19 = 12.76\ \text{Gy}$$

that is, $D_\gamma$ is approximately 17% of the total dose.

5 Measurements in water with a TE ionization chamber (A-150 walls filled with TE-gas) in a p(66) + Be(40) neutron beam yield a reading, corrected for influence quantities, of 18 nC. The chamber has an air-kerma $^{60}$Co $\gamma$-ray calibration coefficient of 50.21 mGy nC$^{-1}$. Determine the mass of TE-gas and the reference absorbed dose in water using the following assumptions and data: $h = 1$, $D_\gamma = 16\%$, $(p_{dis})_{n+\gamma} = 0.987$, $(k_{wall})_{60\,Co} = 0.988$.
*Answer: 0.53 $\mu$g; $\sim$ 1 Gy.*

**Solution:**
The reference absorbed dose to water in a mixed $n+\gamma$ beam is given by Eq. (19.8) in the textbook, that is,

$$(D_w)_{n+\gamma} = (W_{TE\text{-gas}}/e)_{n+\gamma} \frac{(q_{TE\text{-gas}})_{n+\gamma}}{m_{TE\text{-gas}}} (f_{A\text{-}150,TE\text{-gas}})_{n+\gamma} (p_{dis})_{n+\gamma}\, \delta_R (K_{w,A\text{-}150})_{n+\gamma}$$

which requires determining first the mass of TE-gas filling the chamber using Eq. (19.10) with data from Table 19.3:

$$m_{\text{TE-gas}} = \frac{(W_{\text{TE-gas}}/e)_{^{60}\text{Co}}}{N_K} \left[ \frac{s_{\text{A-150,TE-gas}} \ (\bar{\mu}_{\text{en}}/\rho)_{\text{air,A-150}}}{(1 - \bar{g}_{\text{air}}) \ k_{\text{A-150}}} \right]_{^{60}\text{Co}}$$

$$= \frac{29.3 \text{ J C}^{-1}}{50.21 \times 10^6 \text{ Gy C}^{-1}} \ \frac{0.991 \times 0.910}{0.997 \times 0.988} = 5.34 \times 10^{-7} \text{ kg}$$

The next step is to determine $\delta_R$ using Eqs. (19.11) and (19.12) in the textbook, i.e.,

$$k_{\text{TE, w}} = \frac{(W_{\text{TE-gas}})_{^{60}\text{Co}}}{(W_{\text{TE-gas}})_{n+\gamma}} \ \frac{(s_{\text{A-150,TE-gas}})_{^{60}\text{Co}}}{(s_{\text{A-150,TE-gas}})_{n+\gamma}} \ \frac{(K_{\text{w, A-150}})_{^{60}\text{Co}}}{(K_{\text{w, A-150}})_{n+\gamma}}$$

$$= \frac{29.3 \text{ eV}}{31.1 \text{ eV}} \times \frac{0.991}{1.00} \times \frac{1.012}{0.976} = 0.968$$

where $(K_{\text{w, A-150}})_{^{60}\text{Co}}$ has been replaced by $[(\bar{\mu}_{\text{en}}/\rho)_{\text{w, A-150}}]_{^{60}\text{Co}}$. Noting that $D_\gamma = 16\%$ implies $D_n = 84\%$, we have

$$\delta_R = \frac{k_{\text{TE,w}}(D_\gamma + D_n)}{D_\gamma + D_n \ k_{\text{TE,w}}} = 0.995$$

Inserting the values above and those from Table 19.3 into the above expression for the reference absorbed dose results in

$$(D_w)_{n+\gamma} = 31.1 \text{ J C}^{-1} \times \frac{18 \times 10^{-9} \text{ C}}{5.34 \times 10^{-7} \text{ kg}} \times 1.00 \times 0.987 \times 0.995$$

$$\times 0.976 = 1.003 \text{ Gy}$$

6  Assume that in the previous exercise the ionization chamber has instead an absorbed dose-to-water calibration coefficient of $N_{D, \text{w},^{60}\text{Co}} = 53.87 \text{ mGy nC}^{-1}$. What is the absorbed dose to water in the n+γ beam? Would you expect any difference? Assume that $(p_{\text{dis}})_{^{60}\text{Co}} = 0.986$.
   *Answer: 0.997 Gy. Differs by ~ 0.5% from the $N_K$-based $D_w$, which is reasonable considering the uncertainties involved in the $N_K$ and $N_{D,w}$ formalisms.*

**Solution:**
A straightforward application of Eq. (19.14) in the textbook using the $k_{\text{TE,w}}$ and $\delta_R$ values obtained in the previous exercise leads to

$$(D_w)_{n+\gamma} = (q_{\text{TE-gas}})_{n+\gamma} \ N_{D, \text{w},^{60}\text{Co}} \ \frac{(p_{\text{dis}})_{n+\gamma}}{(p_{\text{dis}})_{^{60}\text{Co}}} \ \frac{1}{k_{\text{TE,w}}} \ \delta_R = 18 \text{ nC} \times 53.87$$

$$\times 10^{-3} \text{ Gy nC}^{-1} \times \frac{0.987}{0.986} \times \frac{1}{0.968} \times 0.995 = 0.997 \text{ Gy}$$

The difference from $D_w(N_K)$ in the previous exercise is ~0.5%, which is quite reasonable considering the uncertainties involved in the $N_K$ and $N_{D,w}$ formalisms and in the data used.

7  Write down the necessary expressions for determining the reference absorbed dose to tissue, with measurements made in water, corresponding

to Eq. (19.8) in the textbook. (a) Apply them to ICRU striated muscle using the $N_K$, chamber reading, and assumptions in Exercise 5. (b) Compare with the dose calculation using Eq. (19.15) assuming the neutron fluence ratio to be unity.

*Answer: (a) 0.968; (b) 0.963, that is, a difference of $\sim 0.5\%$.*

**Solution:**

As discussed in the textbook, the different expressions for evaluating $(D_w)_{n+\gamma}$ can be written in terms of $(D_{tissue})_{n+\gamma}$ by replacing the subscript 'w 'by 't '. Obviously, the expression for the mass of the TE-gas filling the chamber is unchanged. This yields

$$(D_t)_{n+\gamma} = (W_{TE\text{-gas}}/e)_{n+\gamma} \frac{(q_{TE\text{-gas}})_{n+\gamma}}{m_{TE\text{-gas}}} (f_{A\text{-}150,TE\text{-gas}})_{n+\gamma} (p_{dis})_{n+\gamma} \, \delta_{R,t} (K_{t,A\text{-}150})_{n+\gamma}$$

and

$$\delta_{R,t} = \frac{k_{TE,t}(D_\gamma + D_n)}{D_\gamma + D_n \, k_{TE,t}}$$

$$k_{TE,t} = \frac{(W_{TE\text{-gas}})_{^{60}Co}}{(W_{TE\text{-gas}})_{n+\gamma}} \frac{(s_{A\text{-}150,TE\text{-gas}})_{^{60}Co}}{(f_{A\text{-}150,TE\text{-gas}})_{n+\gamma}} \frac{(K_{t,\,A\text{-}150})_{^{60}Co}}{(K_{t,\,A\text{-}150})_{n+\gamma}}$$

(a) Using the values in Table 19.3 of the textbook, and replacing $(K_{t,\,A\text{-}150})_{^{60}Co}$ by $[(\bar{\mu}_{en}/\rho)_{t,\,A\text{-}150}]_{^{60}Co}$, it is straightforward to arrive at

$$k_{TE,t} = 0.999$$
$$\delta_{R,t} = 0.999$$
$$(D_t)_{n+\gamma} = 0.968 \text{ Gy}$$

(b) The approximate expression for determining the absorbed dose to tissue from the dose to water is

$$(D_t)_{n+\gamma} \approx (D_w)_{n+\gamma} \frac{\Phi_t}{\Phi_w} \frac{(k_\Phi)_t}{(k_\Phi)_w}$$

From Exercise 5, we have $(D_w)_{n+\gamma} = 1.003$, and here we assume that $\Phi_t/\Phi_w = 1$; the kerma coefficients ratio is identical to $(K_{t,A\text{-}150})_{n+\gamma}$, which for the p(66) + Be(40) spectrum is 0.937. Hence

$$(D_t)_{n+\gamma} \approx 1.003 \times 0.937 = 0.963$$

i.e., a difference from the $(D_t)_{n+\gamma} = 0.968$ Gy above of approximately 0.5%, which is consistent with the approximations involved in both approaches.